My Reference Book

My Reference Book

The McGraw·Hill Companies

UCSMP Elementary Materials Component
Max Bell, Director, UCSMP Elementary Materials Component; Director, *Everyday Mathematics* First Edition
James McBride, Director, *Everyday Mathematics* Second Edition
Andy Isaacs, Director, *Everyday Mathematics* Third Edition
Amy Dillard, Associate Director, *Everyday Mathematics* Third Edition

Authors
Mary Ellen Dairyko, Rachel Malpass McCall, Cheryl G. Moran

Technical Art
Diana Barrie

Technology Advisor
James Flanders

Contributors
Cynthia Annorh, Robert Hartfield, James McBride, Cindy Pauletti, Kara Stalzer, Michael Wilson

---

www.WrightGroup.com

Printed in the United States of America.

Send all inquiries to:
Wright Group/McGraw-Hill
P.O. Box 812960
Chicago, IL 60681

ISBN 0-07-604537-4

16 17 18 WC 12 11 10

The McGraw-Hill Companies

Numbers and Counting 1

Operations and Computation 21

Contents

Games 119

Calculators 159

Index 166

Dear Children,

A reference book is a book that helps people find information. Some other reference books are dictionaries, encyclopedias, and cookbooks. *My Reference Book* can help you find out more about the mathematics you learn in class. You can read this reference book with your teacher, your family, and your classmates.

You will find lots of math in this book.
Some of the things you will find are:

- counts
- shapes
- clocks
- money
- measures
- fractions
- graphs
- patterns
- number stories
- fact families
- math tools
- games

You will also find some big words. A grown-up can help you read these words.

We hope you enjoy this book.

Sincerely,
The Authors

Numbers and Counting

Numbers All Around

Read It Together

Numbers are used in many ways.

Count the shells. How many are there?

- Numbers are used for **counting.**

- Numbers are used as **measures.**

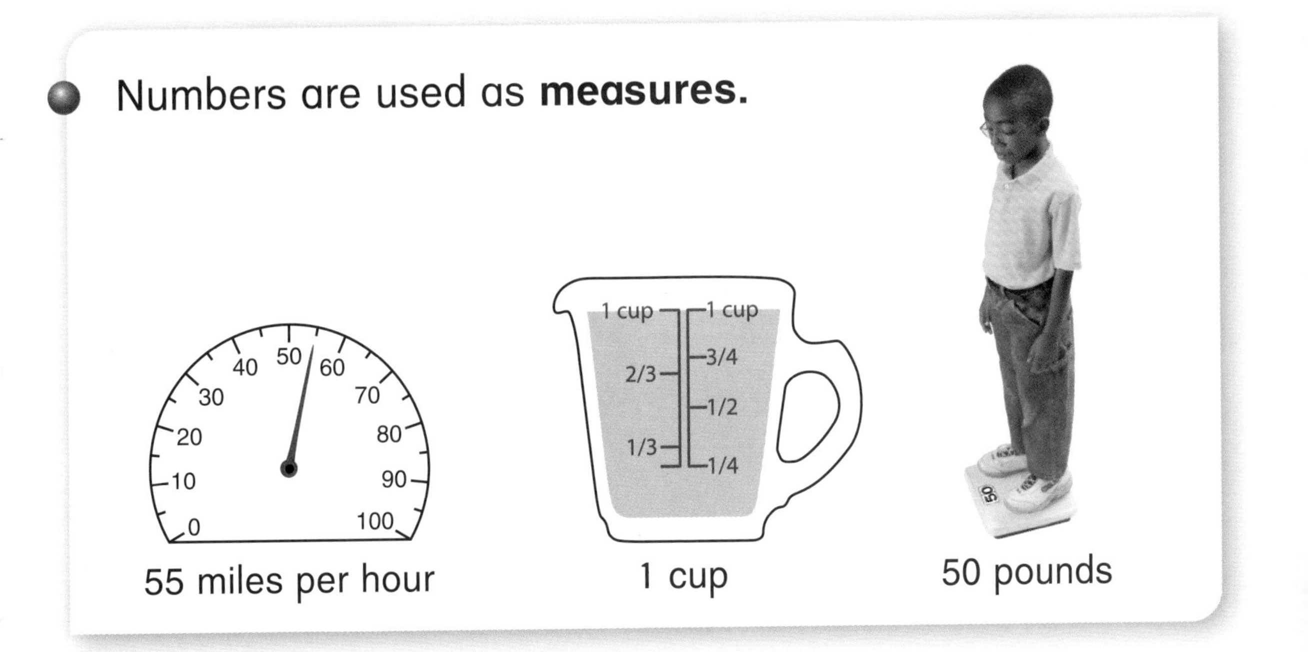

55 miles per hour

1 cup

50 pounds

- Numbers are used to **compare.**

A nickel is worth $\frac{1}{2}$ as much as a dime.

Lily is 5 inches taller than Jake.

Numbers are used as **codes.**

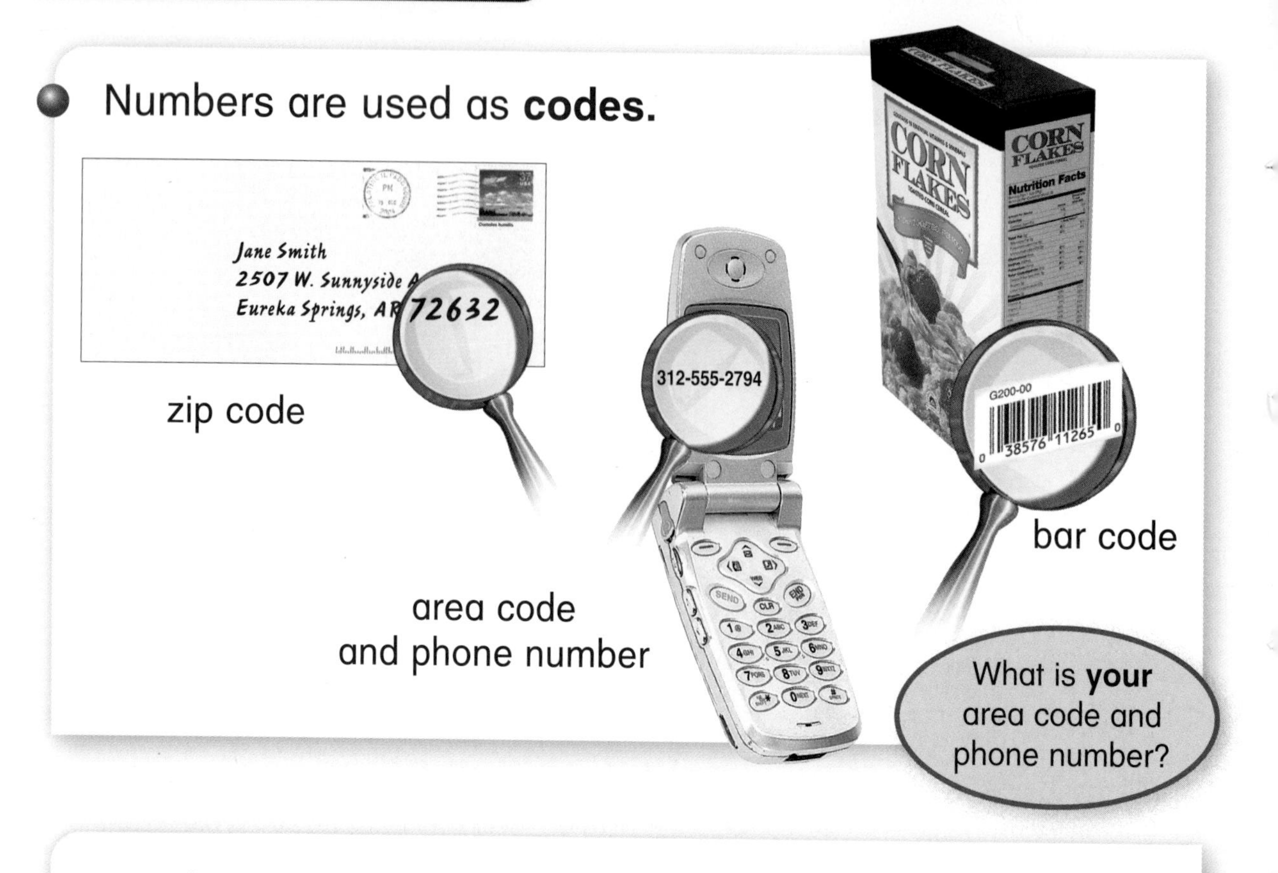

Numbers are used to **show locations.**

We live at
35 Park Street.

The football is on
the 50-yard line.

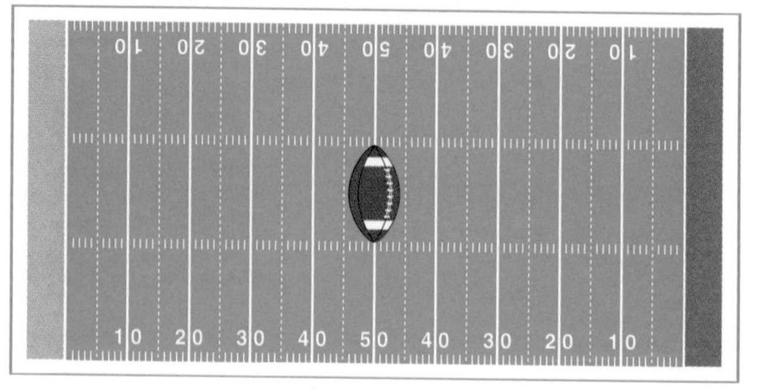

- Numbers are used for **ordering.**

Note

When you use numbers, say what you are counting or measuring. That is called a "unit."

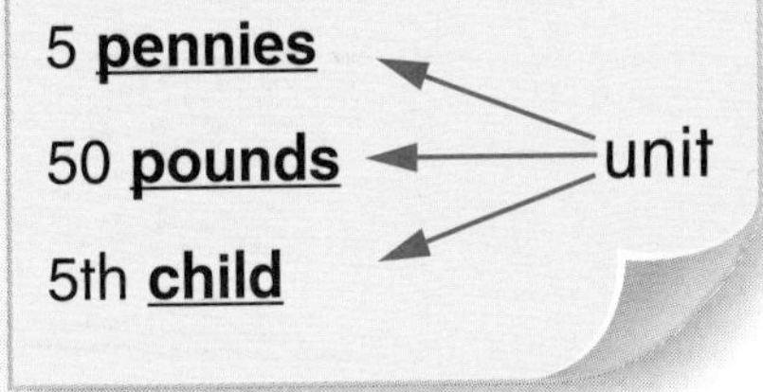

Try It Together

Look around the room. Try to find numbers used in different ways.

Counting Tools

Read It Together

Number lines and **number grids** are tools for counting. To count on a **number line,** think about hopping from one number to another.

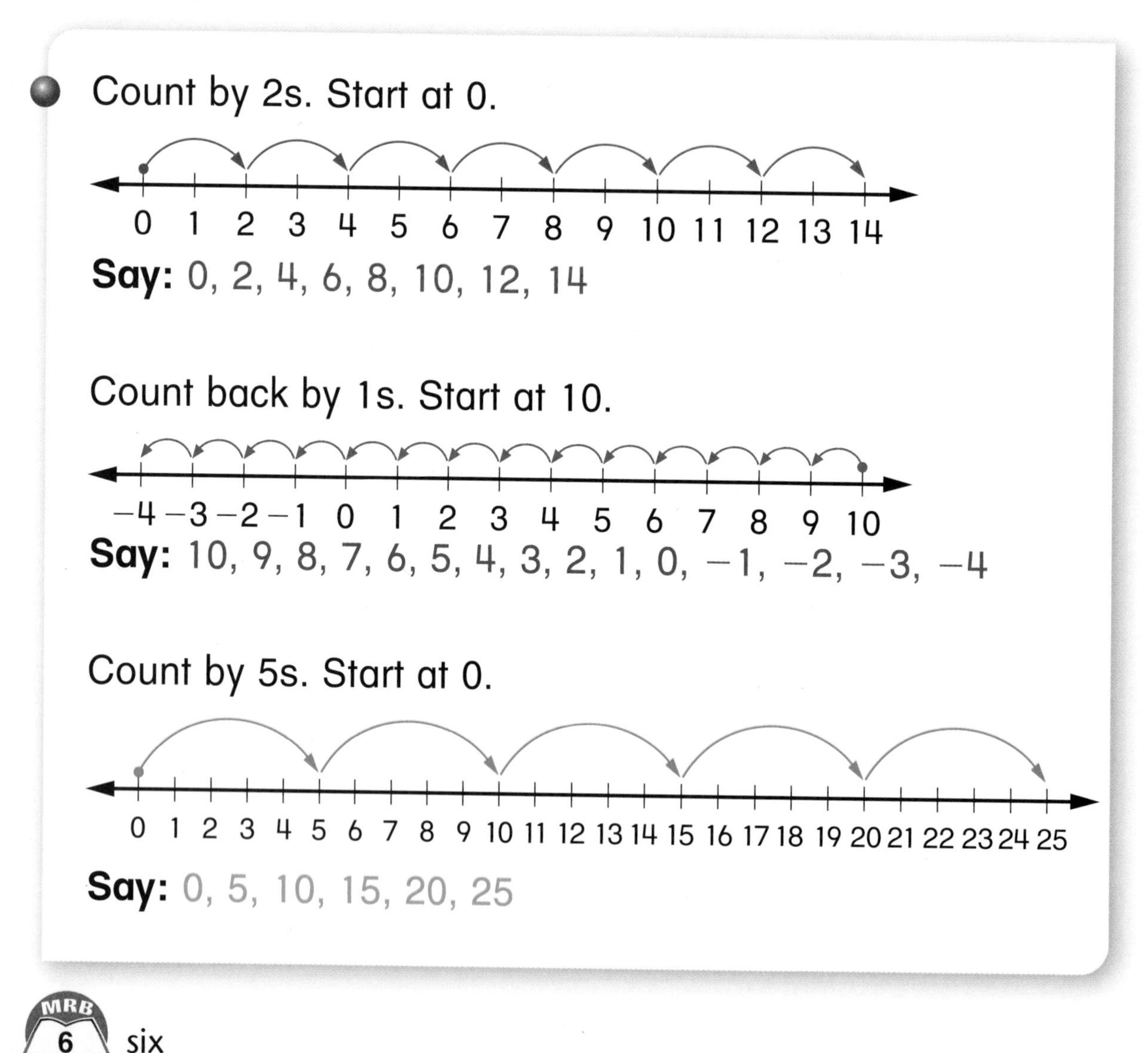

Numbers on a **number grid** are in rows and columns.

−9	−8	−7	−6	−5	−4	−3	−2	−1	**0**
1	2	3	4	5	6	7	8	9	**10**
11	12	13	14	15	16	17	18	19	**20**
21	22	23	24	25	26	27	28	29	**30**
31	32	33	34	35	36	37	38	39	**40**
41	42	43	44	45	46	47	48	49	**50**
51	52	53	54	55	56	57	58	59	**60**
61	62	63	64	65	66	67	68	69	**70**
71	72	73	74	75	76	77	78	79	**80**
81	82	83	84	85	86	87	88	89	**90**
91	92	93	94	95	96	97	98	99	**100**
101	102	103	104	105	106	107	108	109	**110**

Note

To move from row to row, follow the matching colors.

10 → 11

−9	−8	−7	−6	−5	−4	−3	−2	−1	0
1	2	3	4	5	6	7	8	9	10
11	12	13	14	15	16	17	18	19	20
21	22	23	24	25	26	27	28	29	30
31	32	33	34	35	36	37	38	39	40
41	42	43	44	45	46	47	48	49	50
51	52	53	54	55	56	57	58	59	60
61	62	63	64	65	66	67	68	69	70
71	72	73	74	75	76	77	78	79	80
81	82	83	84	85	86	87	88	89	90
91	92	93	94	95	96	97	98	99	100
101	102	103	104	105	106	107	108	109	110

- When you move to the right, numbers get *bigger* by 1.
  For example: 15 is 1 *more* than 14.

  When you move to the left, numbers get *smaller* by 1.
  For example: 23 is 1 *less* than 24.

  When you move down, numbers get *bigger* by 10.
  For example: 37 is 10 *more* than 27.

  When you move up, numbers get *smaller* by 10.
  For example: 43 is 10 *less* than 53.

Comparing Numbers

Read It Together

We use symbols to **compare** numbers.

$25 > 20$
25 **is greater than** 20.

$20 < 25$
20 **is less than** 25.

> **Note**
> The symbol $\neq$ means **is not equal to.**

$20 = 20$
20 **is the same as** 20.
20 **is equal to** 20.

- To help you remember what $>$ and $<$ mean, think about an alligator.

The alligator eats the larger number.

$31 > 18$

Place Value

Read It Together

Numbers can be written using these **digits.**

- The place of a digit in a number tells how much the digit is worth.

thousands	,	hundreds	tens	ones
1	,	3	4	2

The **1** in the **thousands** place is worth **1,000.**
The **3** in the **hundreds** place is worth **300.**
The **4** in the **tens** place is worth **40.**
The **2** in the **ones** place is worth **2.**

The number shown is 1,342.
Say: One thousand, three hundred forty-two.

Base-10 blocks can show numbers.

Base-10 block	Base-10 shorthand	Name	Value
	▪	cube	1
	|	long	10
	□	flat	100
		big cube	1,000

is worth 235.

200 + 30 +5 = 235

is worth 235.

200 + 30 +5 = 235

Try It Together

Use blocks to show 46 and 64. Is the 6 worth more in 46 or 64?

Fractions

Read It Together

A whole object can be divided into **equal parts.**

- a whole clock face

the clock face divided into 4 equal parts

- a whole cracker

the cracker divided into 3 equal parts

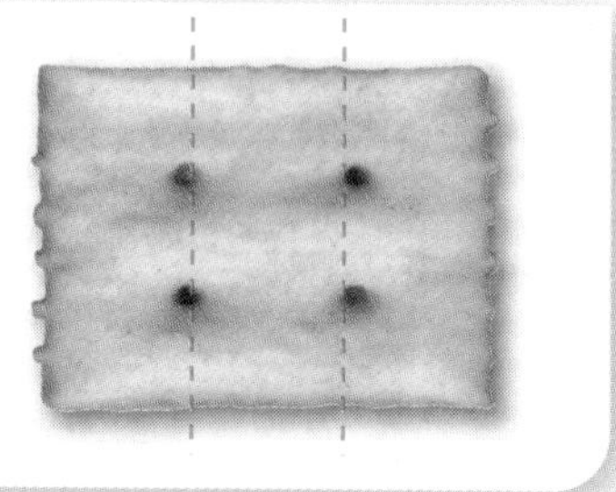

- a whole hexagon

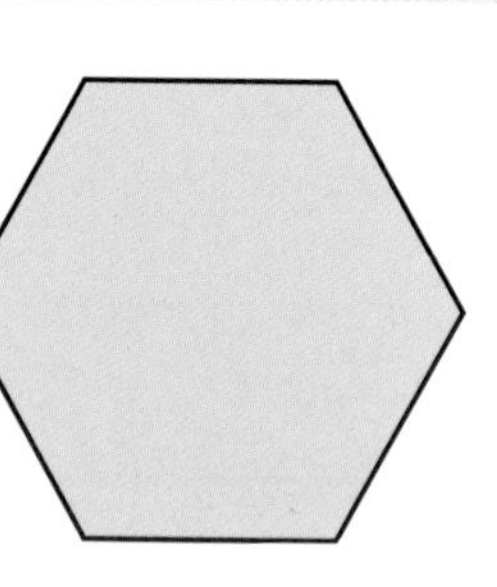

the hexagon divided into 6 equal parts

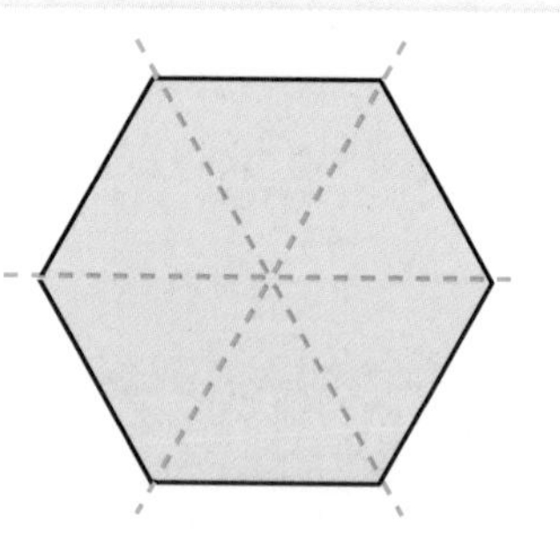

Fractions can name parts of a whole.

The whole clock face is divided into 4 equal parts.

1 part is shaded.

$\frac{1}{4}$ of the clock face is shaded.

$\frac{1}{4}$ ← **numerator** (number of shaded parts)
← **denominator** (number of equal parts)

$\frac{1}{4}$ is a **unit fraction.**
A unit fraction has 1 in the numerator.

- The whole cracker is divided into 3 equal parts.

$\frac{1}{3}$ of the cracker has jelly.
$\frac{2}{3}$ of the cracker does not have jelly.

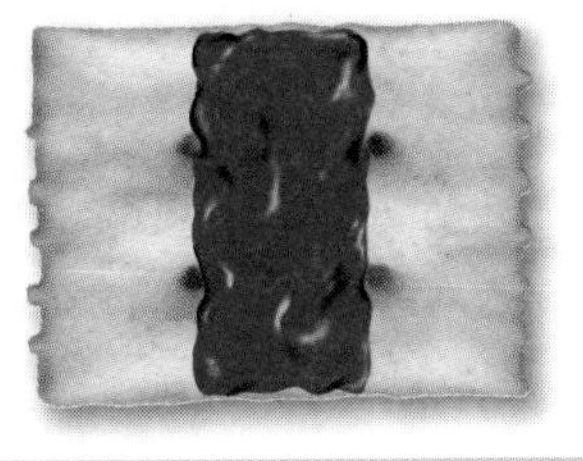

- The whole hexagon is divided into 6 equal parts.

$\frac{4}{6}$ of the hexagon is green.
$\frac{2}{6}$ of the hexagon is yellow.

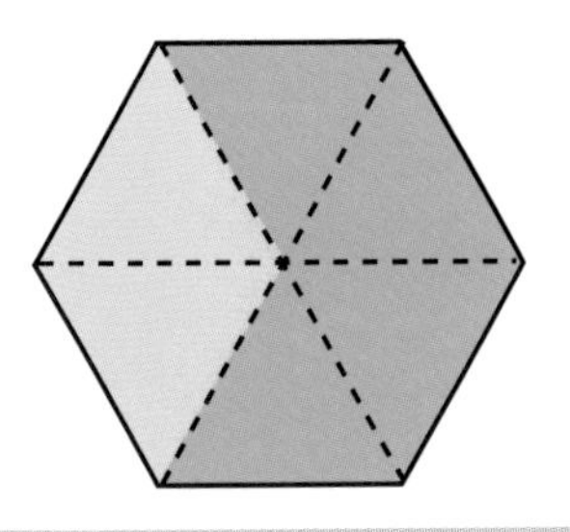

Fractions can name parts of a collection.

- a collection of 6 pennies

3 of the pennies show heads.
$\frac{3}{6}$ of the pennies show heads.

- a collection of 5 flowers

1 of the flowers is red.
$\frac{1}{5}$ of the flowers is red.
← **unit fraction**

Note

When the numerator and the denominator of a fraction are the same, the fraction means "1 whole."

$$\frac{4}{4} = 1$$

Equivalent fractions name the same amount.

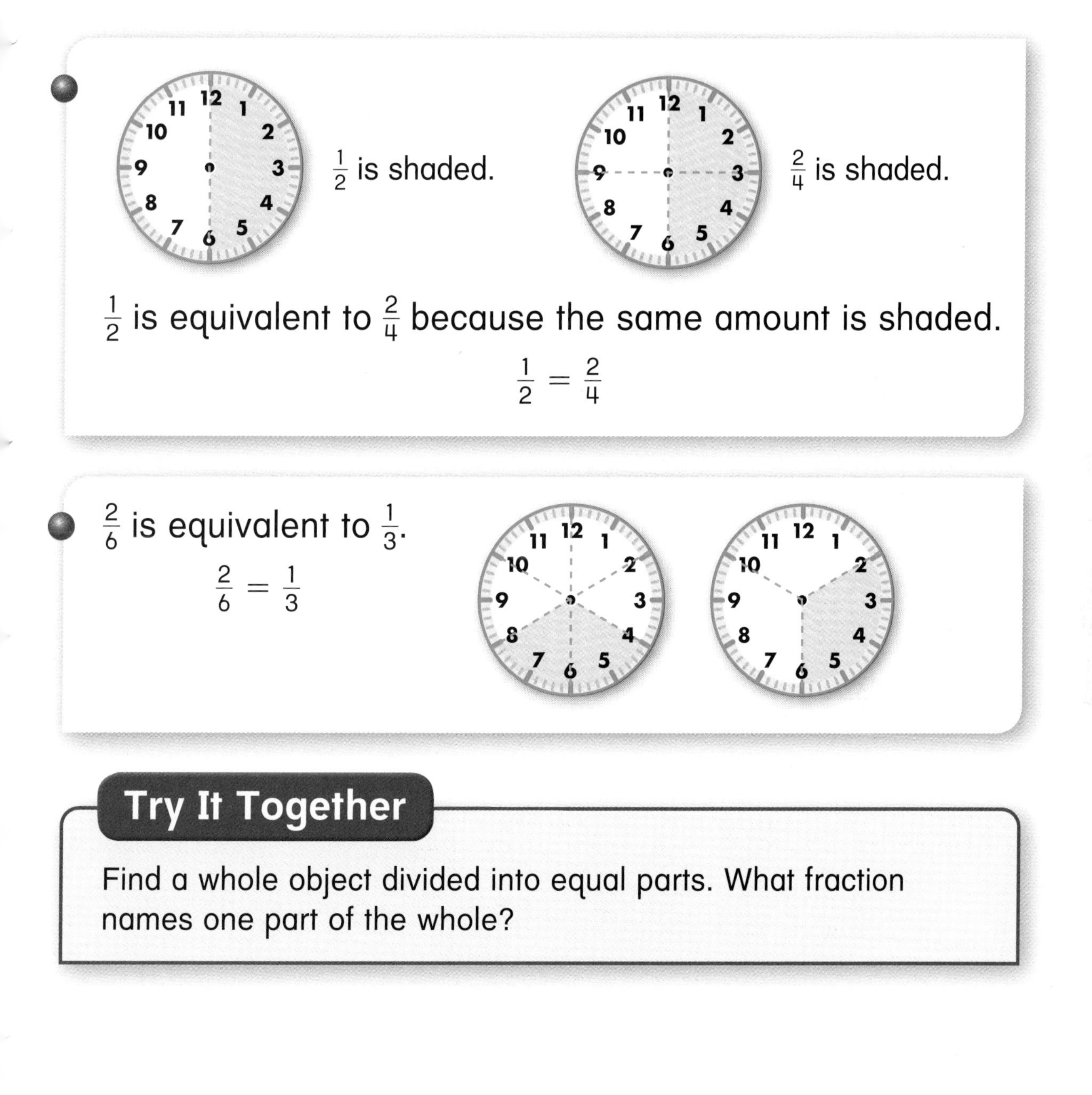

$\frac{1}{2}$ is shaded.

$\frac{2}{4}$ is shaded.

$\frac{1}{2}$ is equivalent to $\frac{2}{4}$ because the same amount is shaded.

$$\frac{1}{2} = \frac{2}{4}$$

$\frac{2}{6}$ is equivalent to $\frac{1}{3}$.

$$\frac{2}{6} = \frac{1}{3}$$

Try It Together

Find a whole object divided into equal parts. What fraction names one part of the whole?

Name-Collection Box

Read It Together

A **name-collection box** is a place to write different names for the same number.

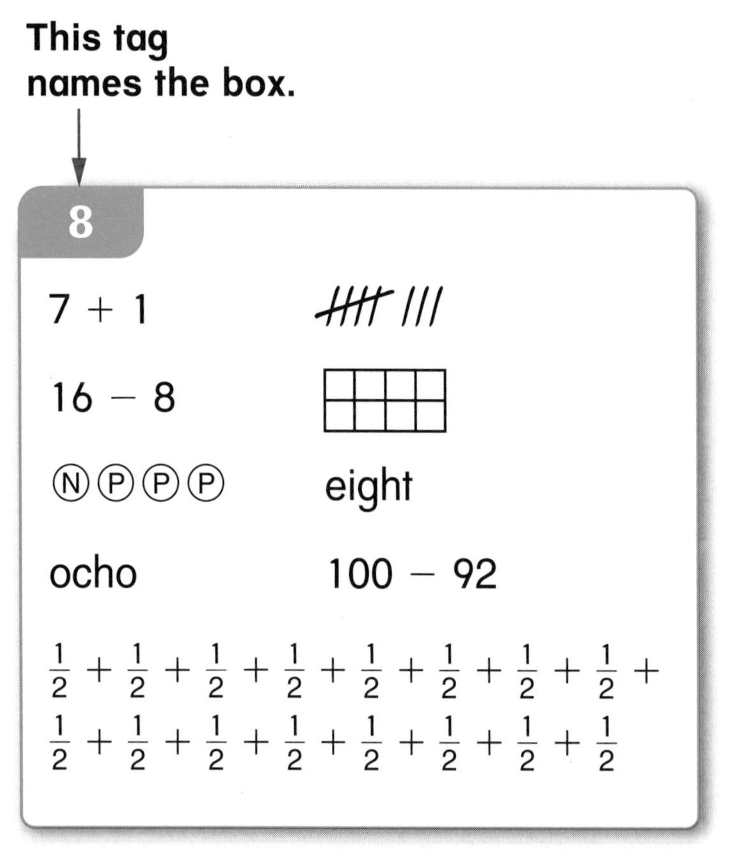

This is a name-collection box for 8.

Numbers All Around: Counting

Mathematics ... Every Day

Counting is an important part of our lives. People count every day.

An umpire counts the number of strikes and balls pitched to each batter.

Swimmers count their strokes so they know when to turn.

Herpetologists study reptiles and amphibians. They count the growth rings of box turtles to find out about how old they are.

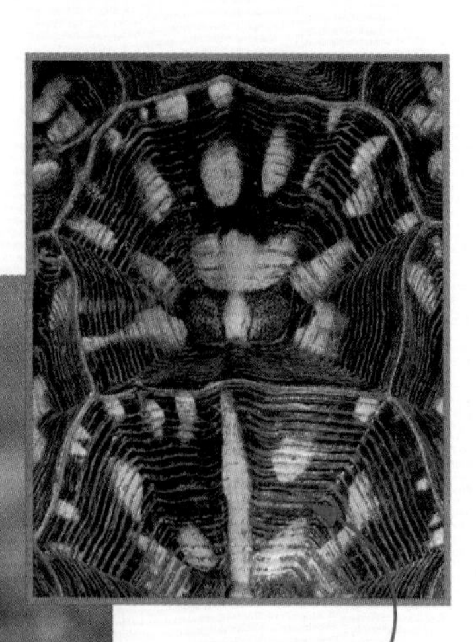

This is a growth ring.

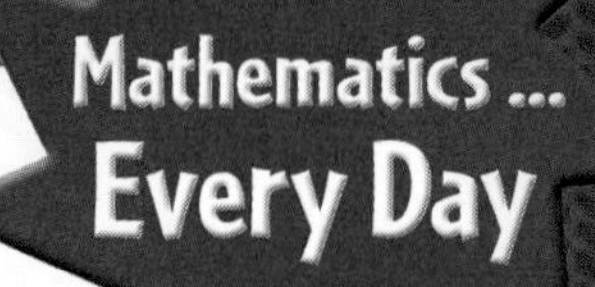

Workers use machines to count the coins made each day at the U.S. Mint. How might you count 10,000 pennies?

Hematologists are blood specialists. They count red blood cells to see how healthy a person is.

Wildlife researchers have to estimate the number of birds in a large flock. Why might wildlife researchers need to count birds? ▼

What things can *you* count?

Operations and Computation

Addition and Subtraction

Read It Together

We use words and symbols for addition and subtraction.

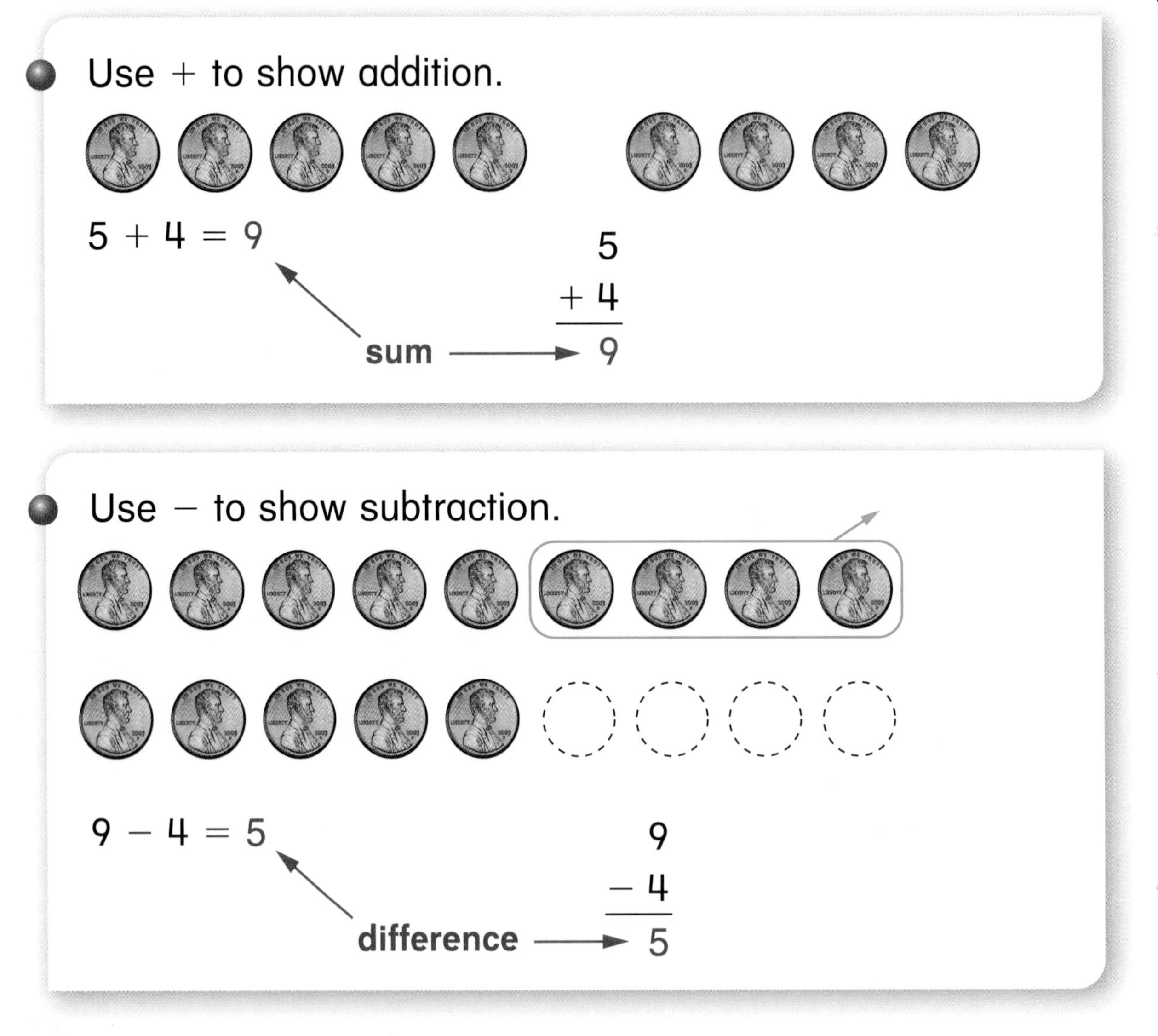

- Use + to show addition.

$5 + 4 = 9$

$$\begin{array}{r} 5 \\ +\ 4 \\ \hline 9 \end{array}$$

sum → 9

- Use − to show subtraction.

$9 - 4 = 5$

$$\begin{array}{r} 9 \\ -\ 4 \\ \hline 5 \end{array}$$

difference → 5

A **number line** can help you find sums.

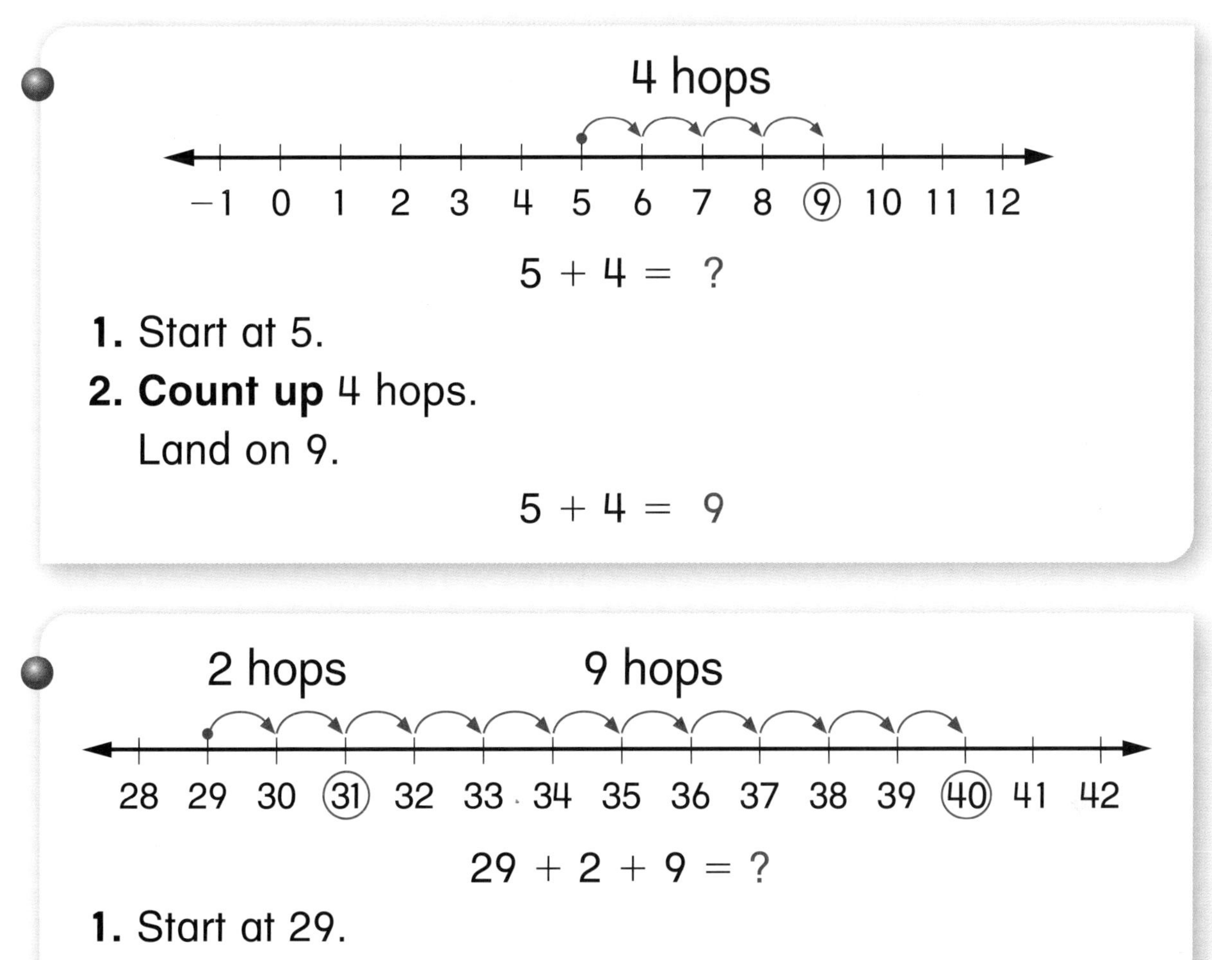

$5 + 4 = ?$

1. Start at 5.
2. **Count up** 4 hops.
   Land on 9.

$5 + 4 = 9$

$29 + 2 + 9 = ?$

1. Start at 29.
2. **Count up** 2 hops.
   Land on 31.
3. **Count up** 9 more hops.
   Land on 40.

$29 + 2 + 9 = 40$

A number line can help you find differences.

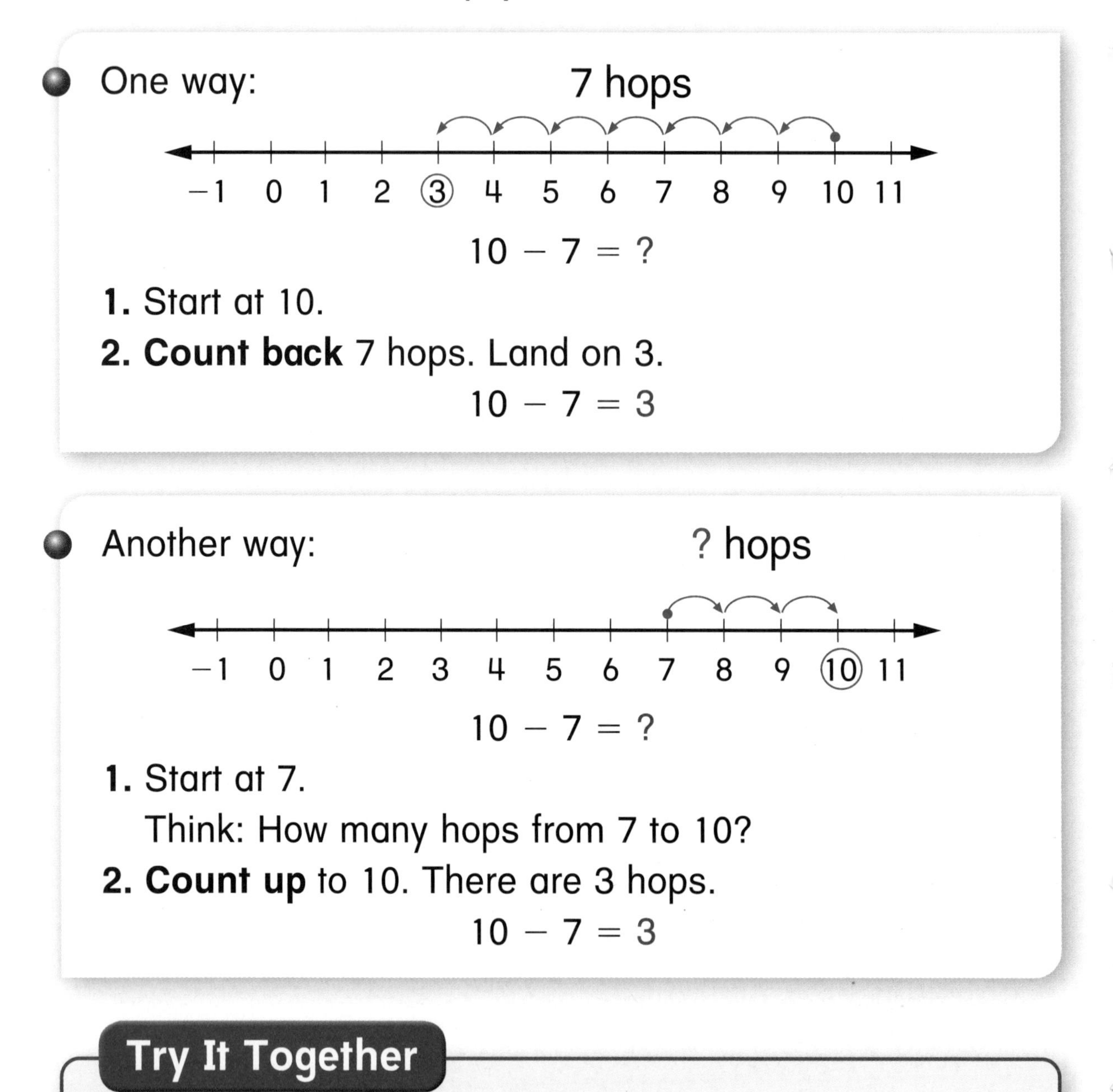

- One way:

$10 - 7 = ?$

1. Start at 10.
2. **Count back** 7 hops. Land on 3.

$10 - 7 = 3$

- Another way:

$10 - 7 = ?$

1. Start at 7.
   Think: How many hops from 7 to 10?
2. **Count up** to 10. There are 3 hops.

$10 - 7 = 3$

Try It Together

Add and subtract numbers on a number line.

Fact Families

Read It Together

A **fact family** is a group of related facts that uses the same numbers.

Dominoes can help you find fact families.

$2 + 6 = 8$

$2 + 6 = 8$ $\quad 8 - 6 = 2$

$6 + 2 = 8$ $\quad 8 - 2 = 6$

This is the fact family for 2, 6, and 8.

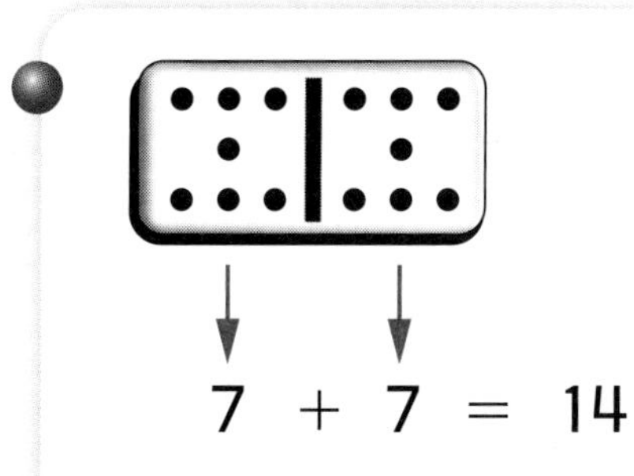

$7 + 7 = 14$

$7 + 7 = 14$ $\quad 14 - 7 = 7$

This **doubles fact family** uses 7, 7, and 14.

Note

Doubles fact families have only two facts instead of four.

You can use an **Addition/Subtraction Facts Table** to help you find fact families.

+,−	0	1	2	3	4	5	6	7	8	9	10
0	0	1	2	3	4	5	6	7	8	9	10
1	1	2	3	4	5	6	7	8	9	10	11
2	2	3	4	5	6	7	8	9	10	11	12
3	3	4	5	6	7	8	9	10	11	12	13
4	4	5	6	7	8	9	10	11	12	13	14
5	5	6	7	8	9	10	11	12	13	14	15
6	6	7	8	9	10	11	12	13	14	15	16
7	7	8	9	10	11	12	13	14	15	16	17
8	8	9	10	11	12	13	14	15	16	17	18
9	9	10	11	12	13	14	15	16	17	18	19
10	10	11	12	13	14	15	16	17	18	19	20

This is the fact family for 9, 7, and 16.

$9 + 7 = 16$ $16 - 7 = 9$

$7 + 9 = 16$ $16 - 9 = 7$

Fact triangles show the 3 numbers in a fact family.

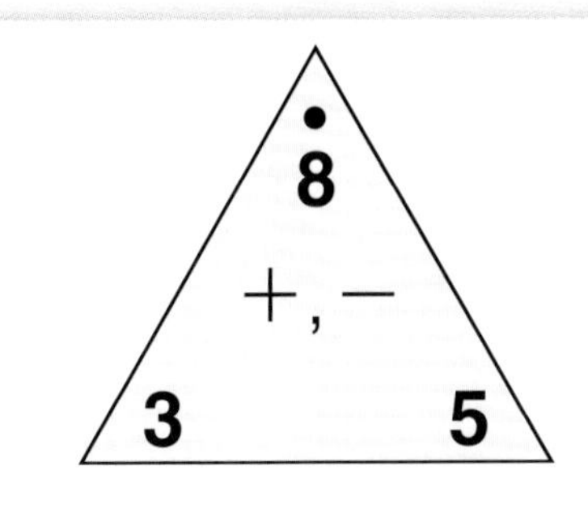

$3 + 5 = 8$ $8 - 5 = 3$
$5 + 3 = 8$ $8 - 3 = 5$

This is the fact family for 8, 5, and 3.

You can use fact triangles to practice facts.

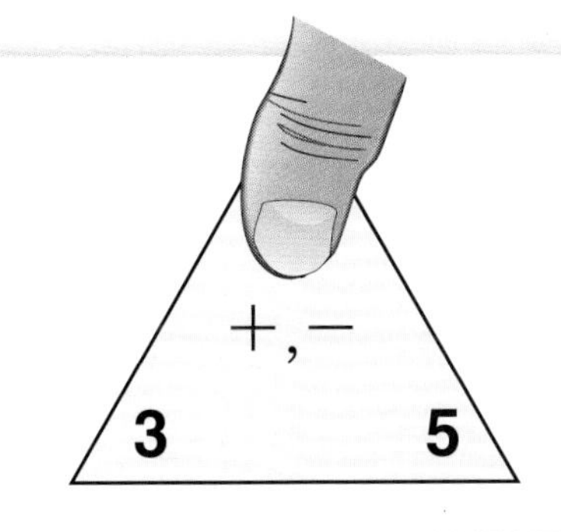

To practice addition, cover the number by the dot.

Cover 8. Think:
$3 + 5 = ?$ $5 + 3 = ?$

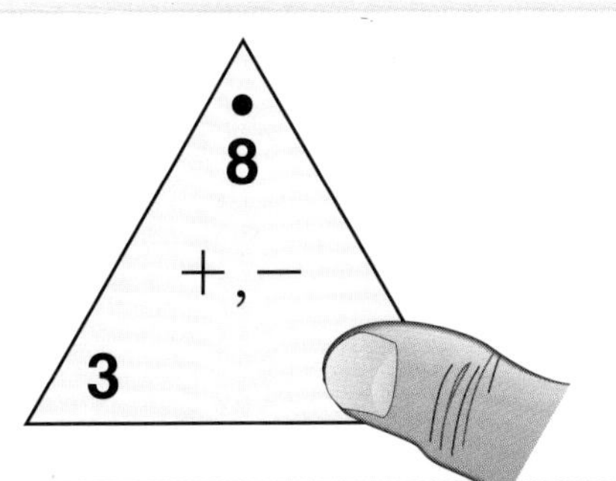

To practice subtraction, cover one of the other numbers.

Cover 5. Think:
$8 - 3 = ?$ $3 + ? = 8$

Try It Together

Use your fact triangles to play *Beat the Calculator.*

Adding Two-Digit Numbers

Read It Together

There are many different ways to add larger numbers.

Try this ⟶ 17 + 25 = ?

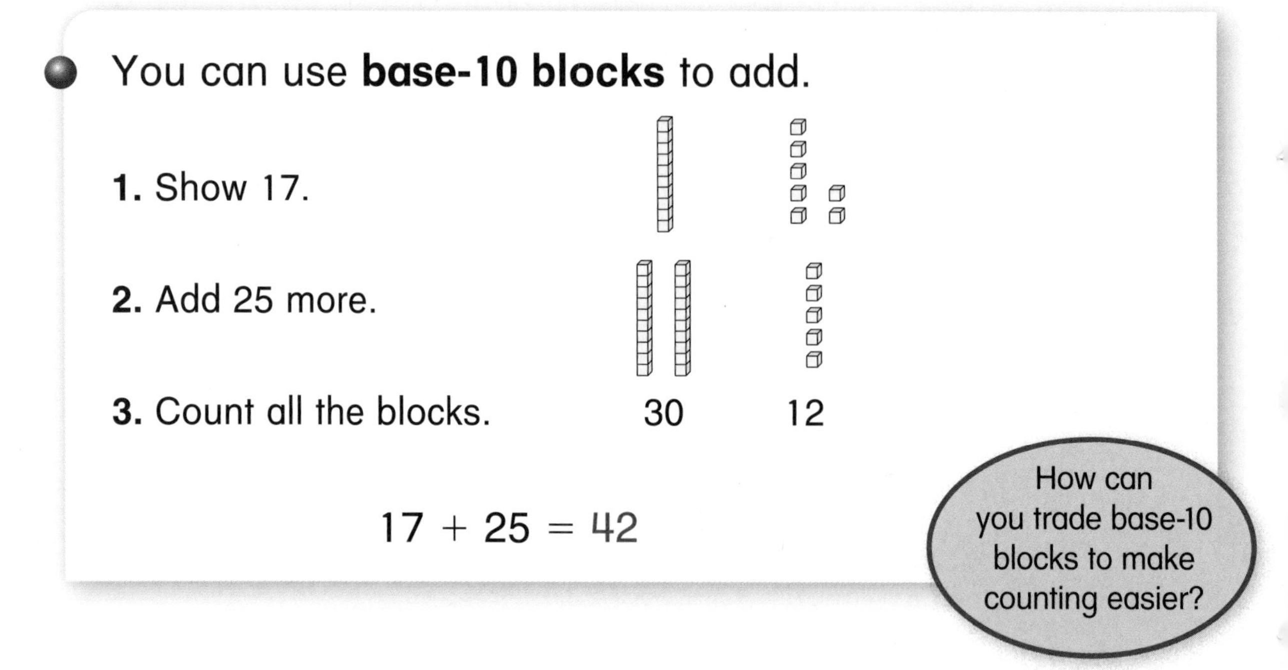

- You can use **base-10 blocks** to add.

1. Show 17.

2. Add 25 more.

3. Count all the blocks. 30 12

17 + 25 = 42

How can you trade base-10 blocks to make counting easier?

You can use a **number grid** to add.

−9	−8	−7	−6	−5	−4	−3	−2	−1	0
1	2	3	4	5	6	7	8	9	10
11	12	13	14	15	16	17	18	19	20
21	22	23	24	25	26	27	28	29	30
31	32	33	34	35	36	37	38	39	40
41	42	43	44	45	46	47	48	49	50
51	52	53	54	55	56	57	58	59	60
61	62	63	64	65	66	67	68	69	70
71	72	73	74	75	76	77	78	79	80
81	82	83	84	85	86	87	88	89	90
91	92	93	94	95	96	97	98	99	100
101	102	103	104	105	106	107	108	109	110

Note

At the end of a row, go to the beginning of the next row and keep counting.

$$17 + 25 = ?$$

1. Start at 17.

2. Add 20.

- Move down 2 rows to 37.

3. Add 5.

- Count 5 more to 42.

$$17 + 25 = 42$$

You can use the **partial-sums addition method** to add.

$17 + 25 = ?$

			10s	1s
			1	**7**
		+	**2**	**5**
1. Add the 10s.	$10 + 20$	=	3	0
2. Add the 1s.	$7 + 5$	=	1	2
3. Add the partial sums.	$30 + 12$	=	**4**	**2**

$17 + 25 = 42$

Try It Together

Add two larger numbers. Show a partner what you did.

Subtracting Two-Digit Numbers

Read It Together

There are many different ways to subtract larger numbers.

Try this ⟶ 43 − 27 = ?

You can use **base-10 blocks** to subtract.

1. Show 43. 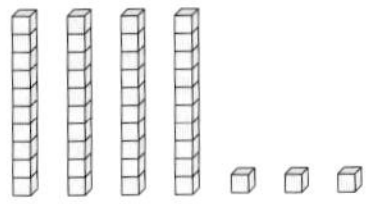
2. Take away 20. 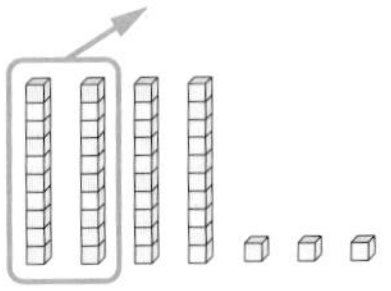
3. Cannot take away 7.
   Trade 1 long for 10 cubes. 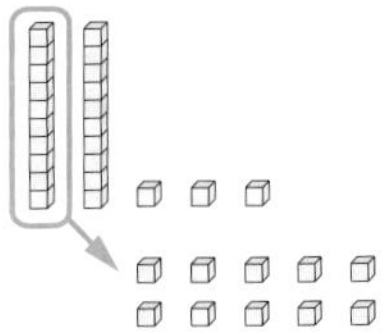
4. Take away 7. 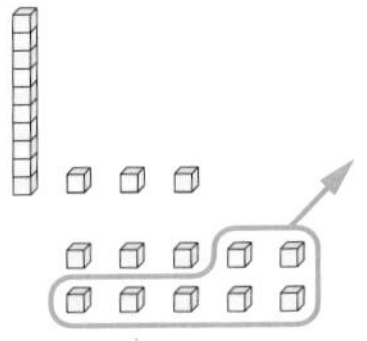
5. 1 long and 6 cubes are left.

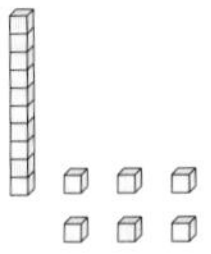

43 − 27 = 16

- You can use a **number grid** to subtract.

−9	−8	−7	−6	−5	−4	−3	−2	−1	0
1	2	3	4	5	6	7	8	9	10
11	12	13	14	15	16	17	18	19	20
21	22	23	24	25	26	27	28	29	30
31	32	33	34	35	36	37	38	39	40
41	42	43	44	45	46	47	48	49	50
51	52	53	54	55	56	57	58	59	60
61	62	63	64	65	66	67	68	69	70
71	72	73	74	75	76	77	78	79	80
81	82	83	84	85	86	87	88	89	90
91	92	93	94	95	96	97	98	99	100
101	102	103	104	105	106	107	108	109	110

$$43 - 27 = ?$$

1. Start at 43.
2. Subtract 20.
   - Move up 2 rows to 23.
3. Subtract 7.
   - Count back 7 to 16.

$$43 - 27 = 16$$

Another way to subtract on a number grid is to count up.

−9	−8	−7	−6	−5	−4	−3	−2	−1	0
1	2	3	4	5	6	7	8	9	10
11	12	13	14	15	16	17	18	19	20
21	22	23	24	25	26	27	28	29	30
31	32	33	34	35	36	37	38	39	40
41	42	43	44	45	46	47	48	49	50
51	52	53	54	55	56	57	58	59	60
61	62	63	64	65	66	67	68	69	70
71	72	73	74	75	76	77	78	79	80
81	82	83	84	85	86	87	88	89	90
91	92	93	94	95	96	97	98	99	100
101	102	103	104	105	106	107	108	109	110

Note

You can count up to find how much change you get back at the store.

$$43 - 27 = ?$$

1. Start at 27.

2. Count up 10.
- Move down 1 row to 37.

3. Count up 6 more.
- Count 6 more to 43.

$$43 - 27 = 16$$

You can use the **trade-first subtraction method** to subtract.

$$43 - 27 = ?$$

You cannot take 7 ones from 3 ones without getting a negative number.

10s	1s
4	**3**
− 2	**7**

1. So trade 1 ten for 10 ones.
4 tens − 1 ten = 3 tens
3 ones + 10 ones = 13 ones

10s	1s
3	**13**
~~4~~	~~3~~
− 2	7

2. Subtract the tens and the ones.

10s	1s
3	**13**
~~4~~	~~3~~
− 2	7
1	**6**

$$43 - 27 = 16$$

Why can you trade 10 ones for 1 ten?

You can use the **counting-up subtraction method** to subtract.

$$43 - 27 = ?$$

1. Start at 27.
Count up to the nearest ten.

2. Count up by tens to 40.

3. Count up to 43.

$$\begin{array}{r} 27 \\ + \textcircled{3} \\ \hline 30 \\ + \textcircled{10} \\ \hline 40 \\ + \textcircled{3} \\ \hline \mathbf{43} \end{array}$$

Circle each number that you count up.

4. Add the numbers you circled.

$$\begin{array}{r} 3 \\ 10 \\ + \ 3 \\ \hline 16 \end{array}$$

$3 + 10 + 3 = 16$

Think:

27 $\xrightarrow{+3}$ 30 $\xrightarrow{+10}$ 40 $\xrightarrow{+3}$ 43

$3 + 10 + 3 = 16$

$$43 - 27 = 16$$

Try It Together

Subtract two larger numbers. Show a partner what you did.

Equal Groups and Equal Shares

Read It Together

Equal groups have the same number of things.

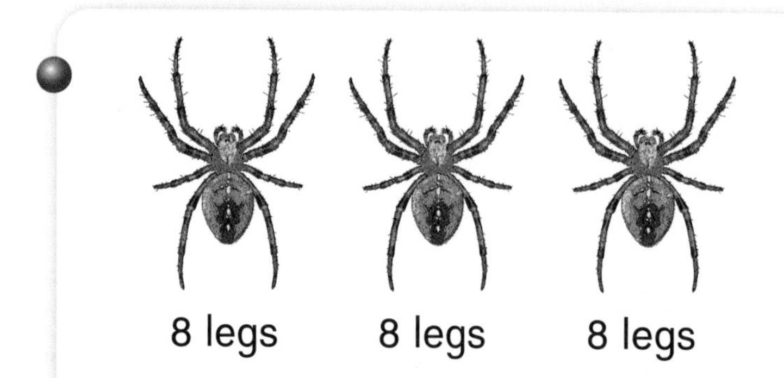

There are three spiders.
Each spider has 8 legs.
There are 24 legs in all.

Sometimes you have things that you want to share equally.

When you make **equal shares,** each group has the same number of objects.

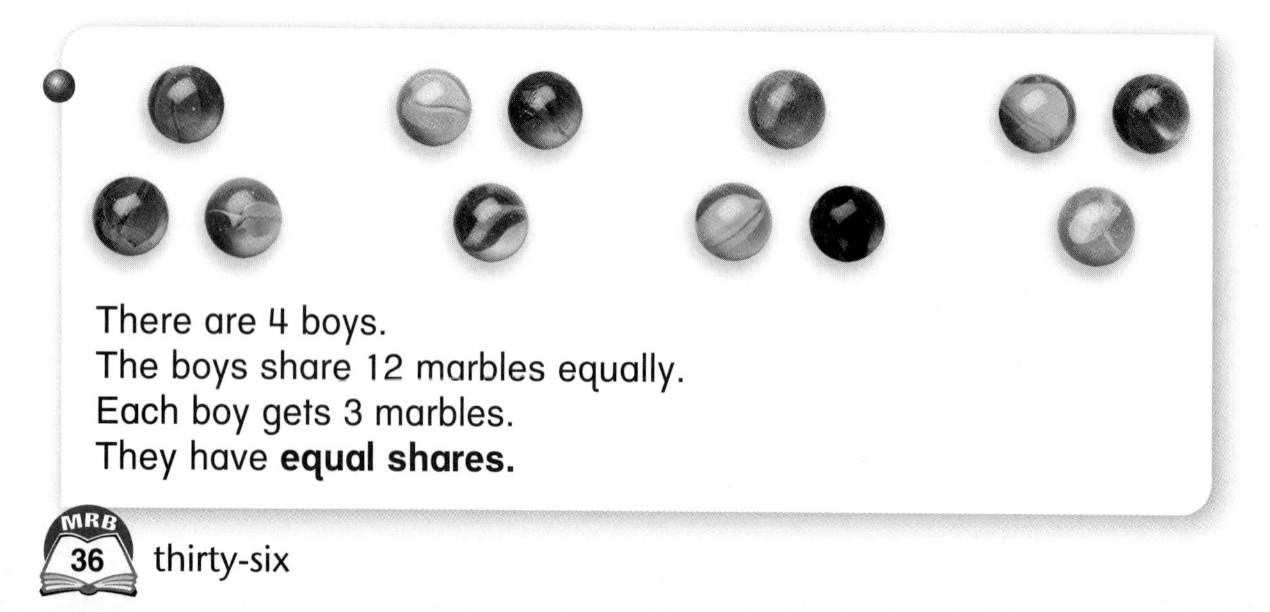

There are 4 boys.
The boys share 12 marbles equally.
Each boy gets 3 marbles.
They have **equal shares.**

Multiplication and Division Fact Triangles

Read It Together

Multiplication and division **fact triangles** show the 3 numbers in a fact family.

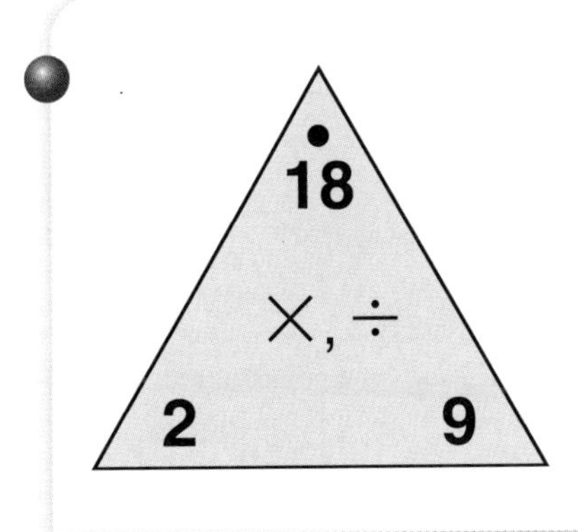

$2 \times 9 = 18$ $18 \div 9 = 2$

$9 \times 2 = 18$ $18 \div 2 = 9$

This is the fact family for 2, 9, and 18.

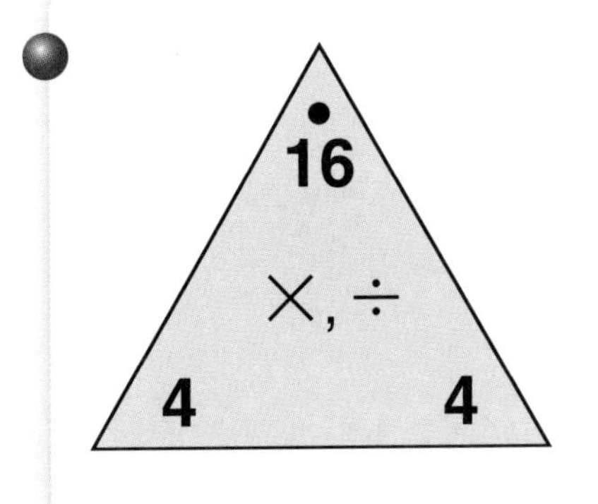

$4 \times 4 = 16$ $16 \div 4 = 4$

This is the fact family for 4, 4, and 16.

You can use fact triangles to practice facts.

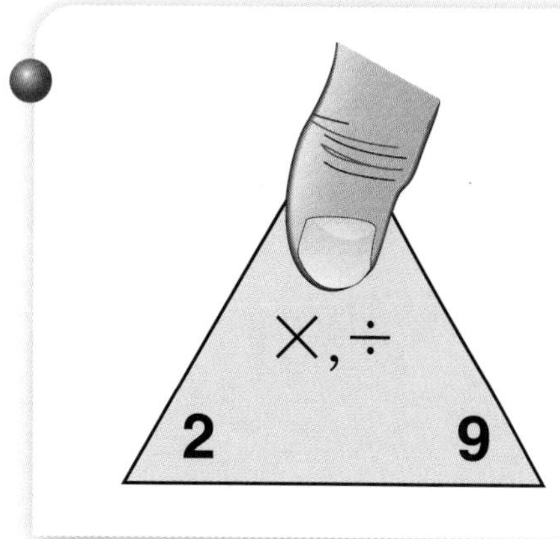

To practice multiplication, cover the number by the dot.

Cover 18. Think:

$2 \times 9 = ?$ $9 \times 2 = ?$

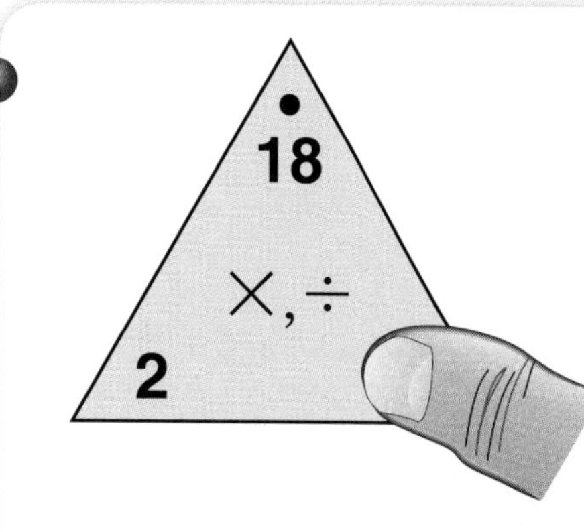

To practice division, cover one of the other numbers.

Cover 9. Think:

$18 \div 2 = ?$ $2 \times ? = 18$

Try It Together

Use your multiplication and division fact triangles to play *Beat the Calculator*.

Data and Chance

Tally Chart and Line Plot

Read It Together

Information someone has collected is called **data.**

- A **tally chart** is one way to organize data.

Teeth Lost in Mr. Alan's Class

Number of Teeth Lost	Number of Children
0	卌
1	卌 //
2	///
3	////
4	/
5	

Note

Remember: Every fifth tally mark goes across a group of four tally marks.

卌

This tally chart shows the number of teeth lost by the children in Mr. Alan's class.

There are 3 tally marks to the right of 2. This means that 3 children lost 2 teeth.

- A **line plot** is another way to organize data.

Teeth Lost in Mr. Alan's Class

Number of Children						
		X				
		X				
	X	X				
	X	X		X		
	X	X	X	X		
	X	X	X	X		
	X	X	X	X	X	
	0	1	2	3	4	5

Number of Teeth Lost

This line plot shows the same data as the tally chart on page 40.

There are 3 Xs above 2.
This means that 3 children lost 2 teeth.

How many teeth have you lost?

Try It Together

How many children did not lose any teeth?

Graphs

Read It Together

A **picture graph** uses a picture or a symbol to show data.

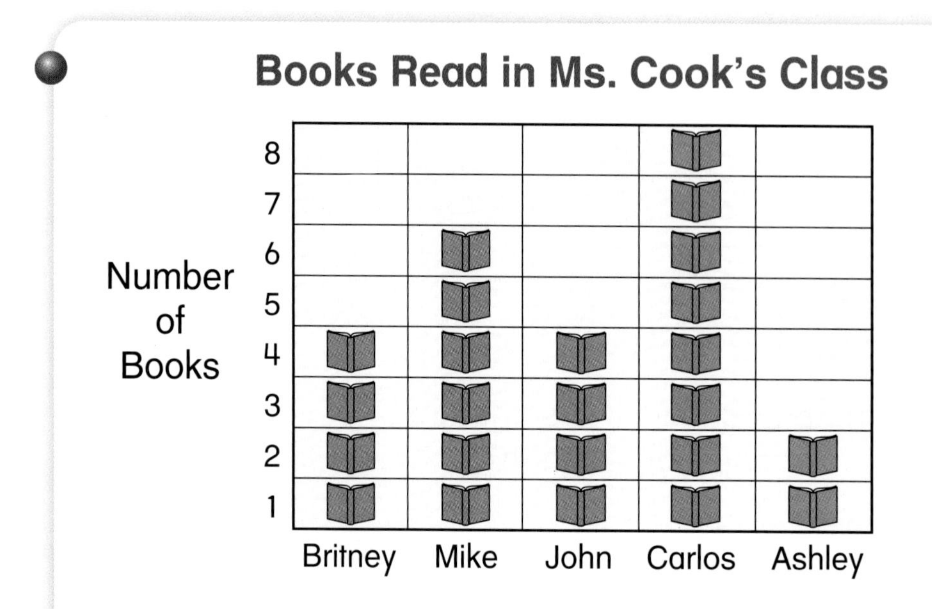

There are 4 above Britney.
This means that Britney read 4 books.

There are 6 above Mike.
This means that Mike read 6 books.

Mike read more books than Britney.

A **pictograph** also uses pictures or symbols to show data.

The **key** on a pictograph tells what each picture stands for.

Books Read in Ms. Cook's Class

Britney	📖 📖
Mike	📖 📖 📖
John	📖 📖
Carlos	📖 📖 📖 📖
Ashley	📖

Key: Each 📖 = 2 books.

This pictograph shows the same data as the picture graph on page 42.

The key on this pictograph tells us that each 📖 stands for 2 books read.

There are 2 📖 to the right of Britney. This means that Britney read 4 books.

There are 3 📖 to the right of Mike. This means that Mike read 6 books.

Carlos read the most books.

A **bar graph** uses bars to show data.

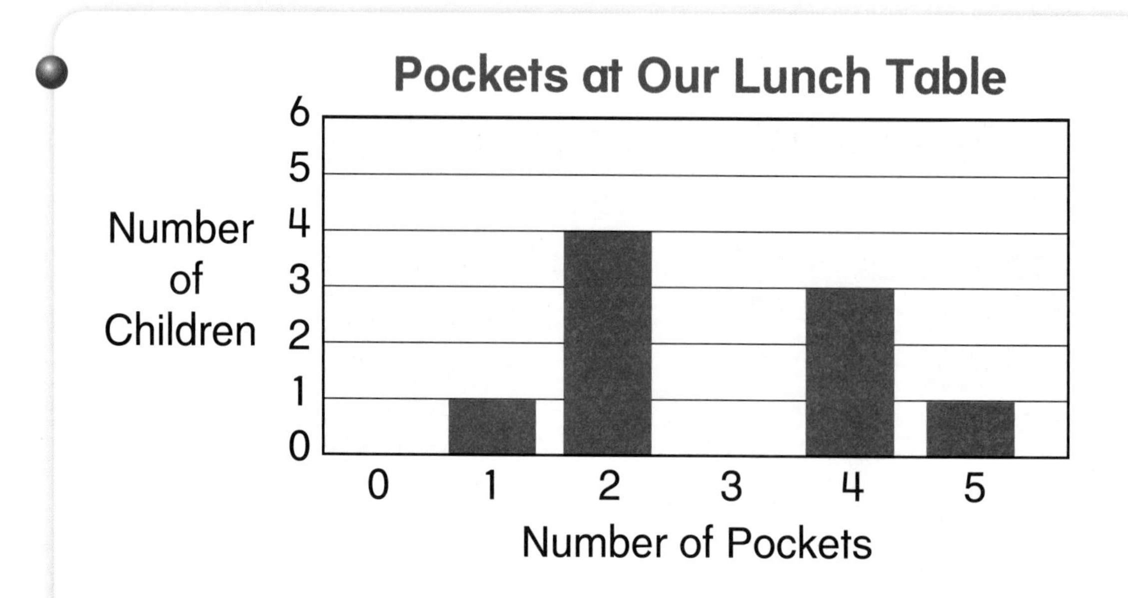

The bar above 5 shows that 1 child has 5 pockets.

The bar above 4 is taller. This shows that more children have 4 pockets than 5 pockets.

There is no bar above 3. This shows that no children have 3 pockets.

Try It Together

How many children are at the lunch table?

Describing Data

Read It Together

Here are some numbers that describe data.

- The **mode** is the number that occurs most often.
The mode of pockets at the lunch table is **2.**
More children have 2 pockets than any other number of pockets.

The **maximum** is the largest number.
The maximum number of pockets at the lunch table is **5.**
One child has 5 pockets.

The **minimum** is the smallest number.
The minimum number of pockets at the lunch table is **1.**
One child has 1 pocket.

The **range** is the difference between the largest and smallest numbers. Subtract the minimum from the maximum to find the **range.**

maximum	5 pockets
minimum	− 1 pocket
range	**4 pockets**

The range of the number of pockets is **4.**

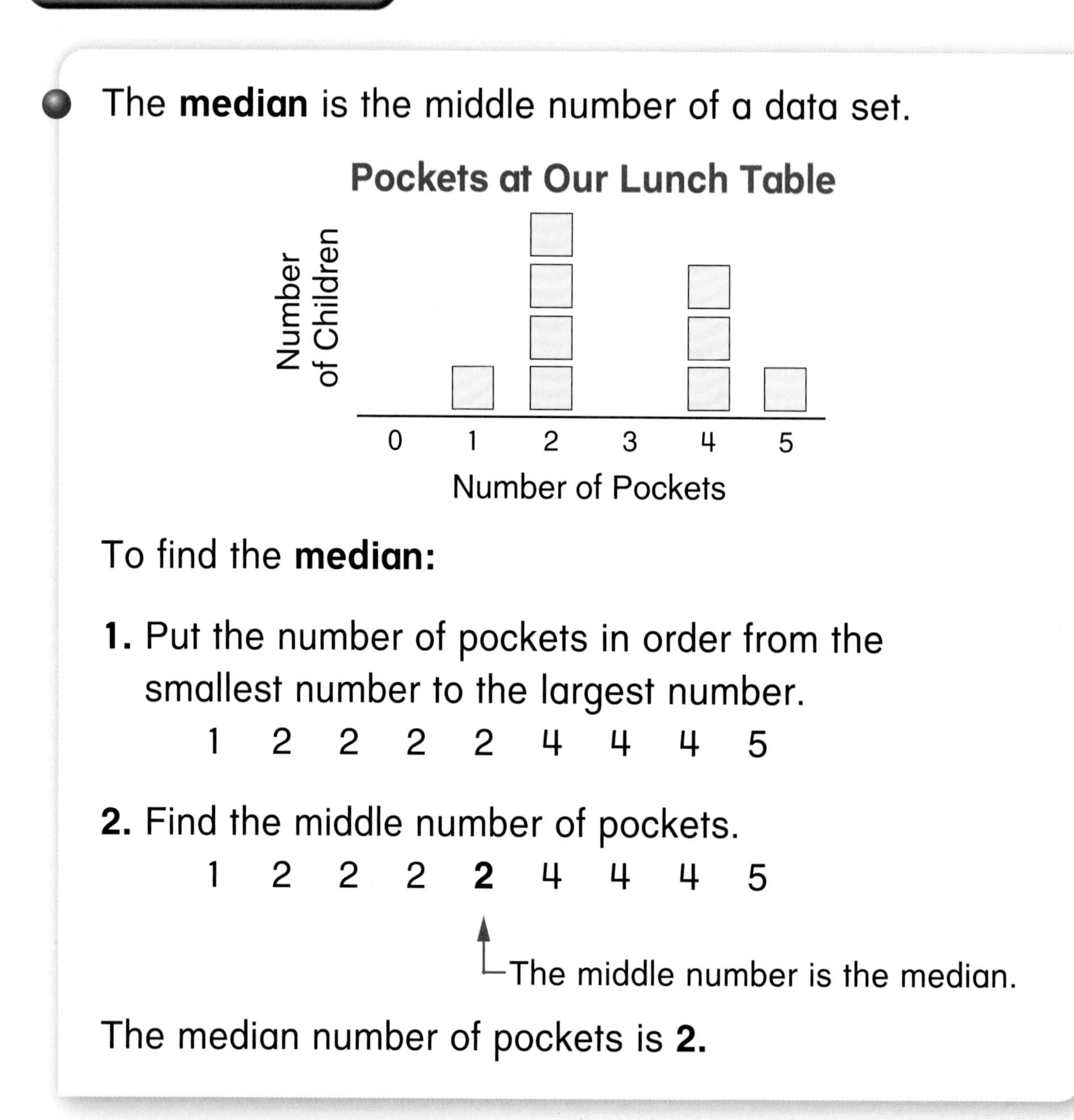

The **median** is the middle number of a data set.

To find the **median:**

1. Put the number of pockets in order from the smallest number to the largest number.

   1 2 2 2 2 4 4 4 5

2. Find the middle number of pockets.

   1 2 2 2 **2** 4 4 4 5

   The middle number is the median.

The median number of pockets is **2.**

Try It Together

Ask 5 people how many pockets they have. Organize the data you collect. Then find the mode, range, and median.

Chance

Read It Together

When you spin a color spinner, you might spin any of the colors. You can **predict** what color you will spin, but you might be wrong.

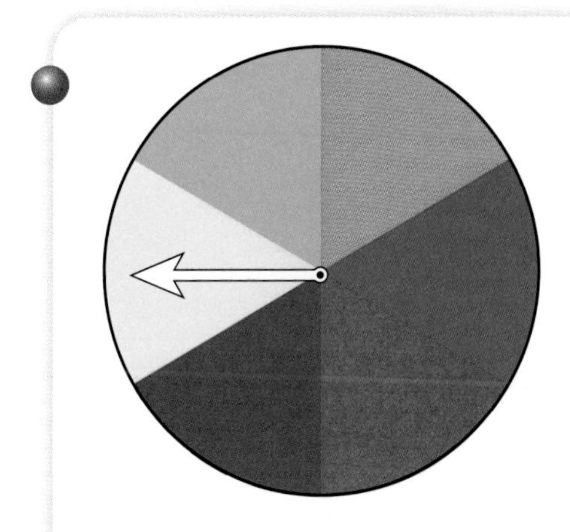

The spinner has an **equal chance** of landing on yellow, green, orange, blue, red, or purple.

The spinner has **no chance** of landing on pink because there is no pink on the spinner.

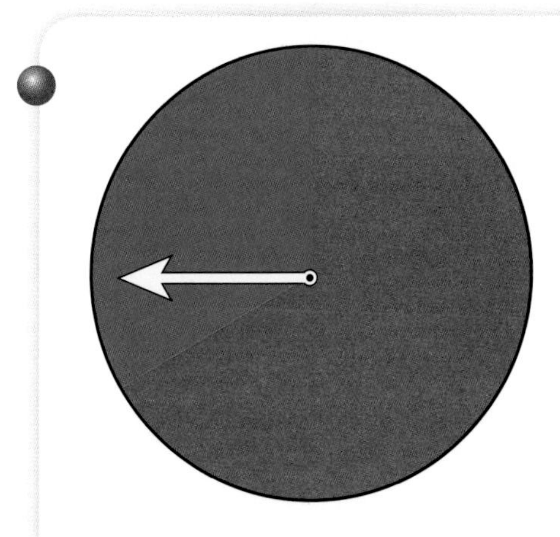

It is **certain** that the spinner will land on either red or blue.

The chance of landing on red is greater than the chance of landing on blue.

What is most likely to happen?

- Suppose you close your eyes, reach in the bag, and grab a block.

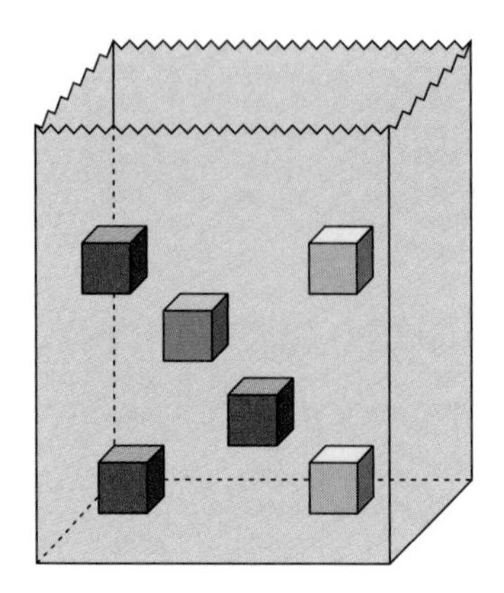

3 red blocks
2 yellow blocks
1 blue block

You are **most likely** to grab a red block because there are more red than either blue or yellow blocks.

You are **least likely** to grab a blue block because there are fewer blue than either red or yellow blocks.

Try It Together

Put 1 red crayon, 2 yellow crayons, and 3 blue crayons in a bag. Which crayon are you most likely to grab? Which crayon are you least likely to grab? Then close your eyes, reach in the bag, and grab a crayon. What might happen if you did it again? What if you did it 10 times? 100 times?

Geometry

Points, Line Segments, and Lines

Read It Together

A **point** is a location in space.
A point is like a dot.
We show a point like this: $\overset{\bullet}{A}$
It is called point *A*.

Line segments and **lines** are made of points.

Shape	Name of Shape	Symbol for Shape
endpoints; C, D	This is a **line segment.** It has 2 **endpoints.** It is called line segment *CD* or line segment *DC*.	$\overline{CD}$ or $\overline{DC}$
F, E	This is a **line.** It is called line *EF* or line *FE*. The arrowheads show that the line goes on and on in both directions.	$\overleftrightarrow{EF}$ or $\overleftrightarrow{FE}$

Parallel lines never meet. They are always the same distance apart.

Parallel line segments are parts of parallel lines.

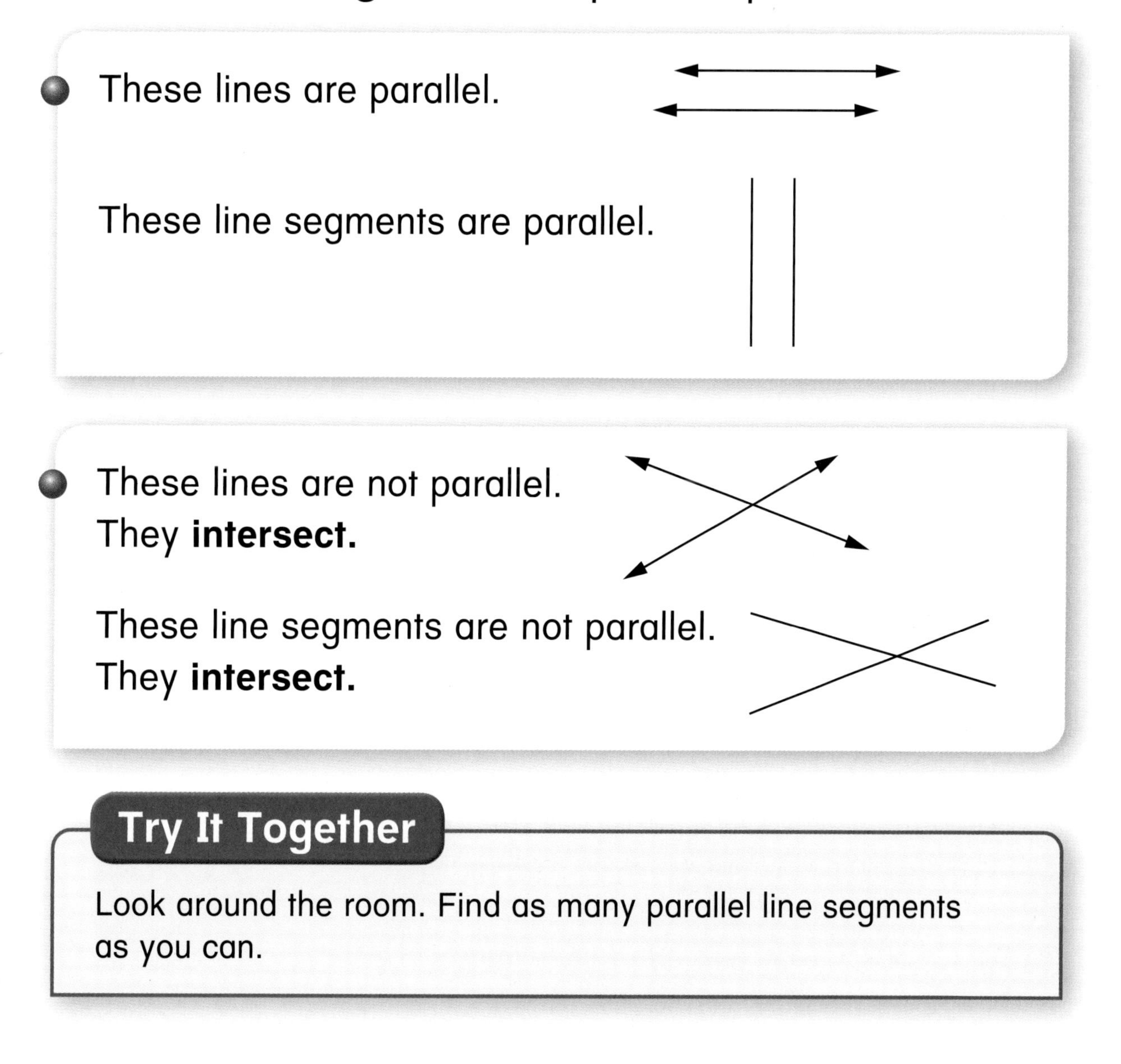

- These lines are parallel.

  These line segments are parallel.

- These lines are not parallel.
  They **intersect.**

  These line segments are not parallel.
  They **intersect.**

Try It Together

Look around the room. Find as many parallel line segments as you can.

2-Dimensional Shapes

Read It Together

A plane shape is flat. Plane shapes are also called **2-dimensional.** You can draw them on paper.

- This 2-dimensional shape is a **polygon.** Polygons are made of line segments.

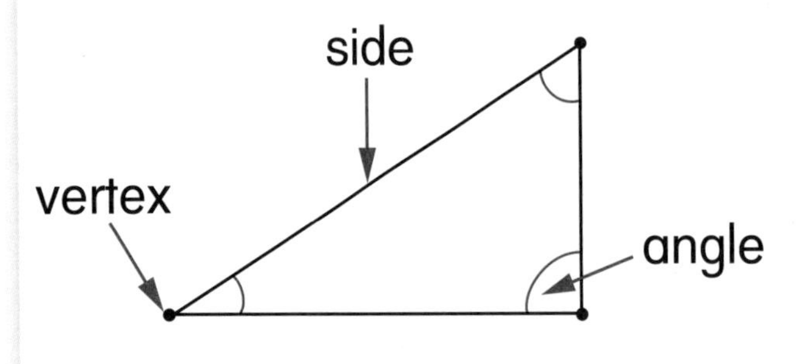

This polygon has 3 **sides,** 3 **angles,** and 3 **vertices.**

Note

A point where two sides meet is called a **vertex.** The plural of *vertex* is *vertices.*

- These 2-dimensional shapes are also polygons.

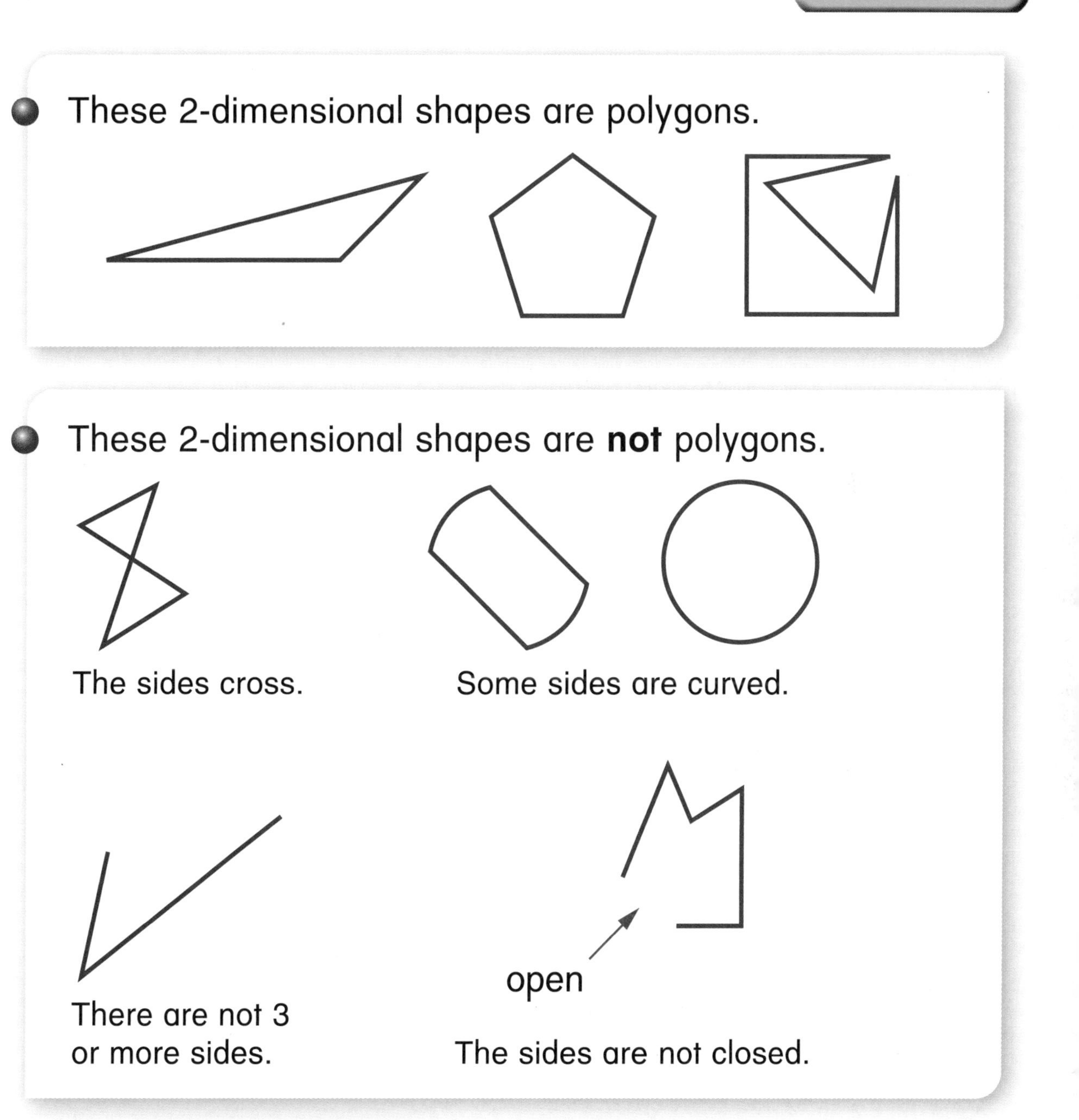

Note

All polygons are closed shapes but not all closed shapes are polygons.

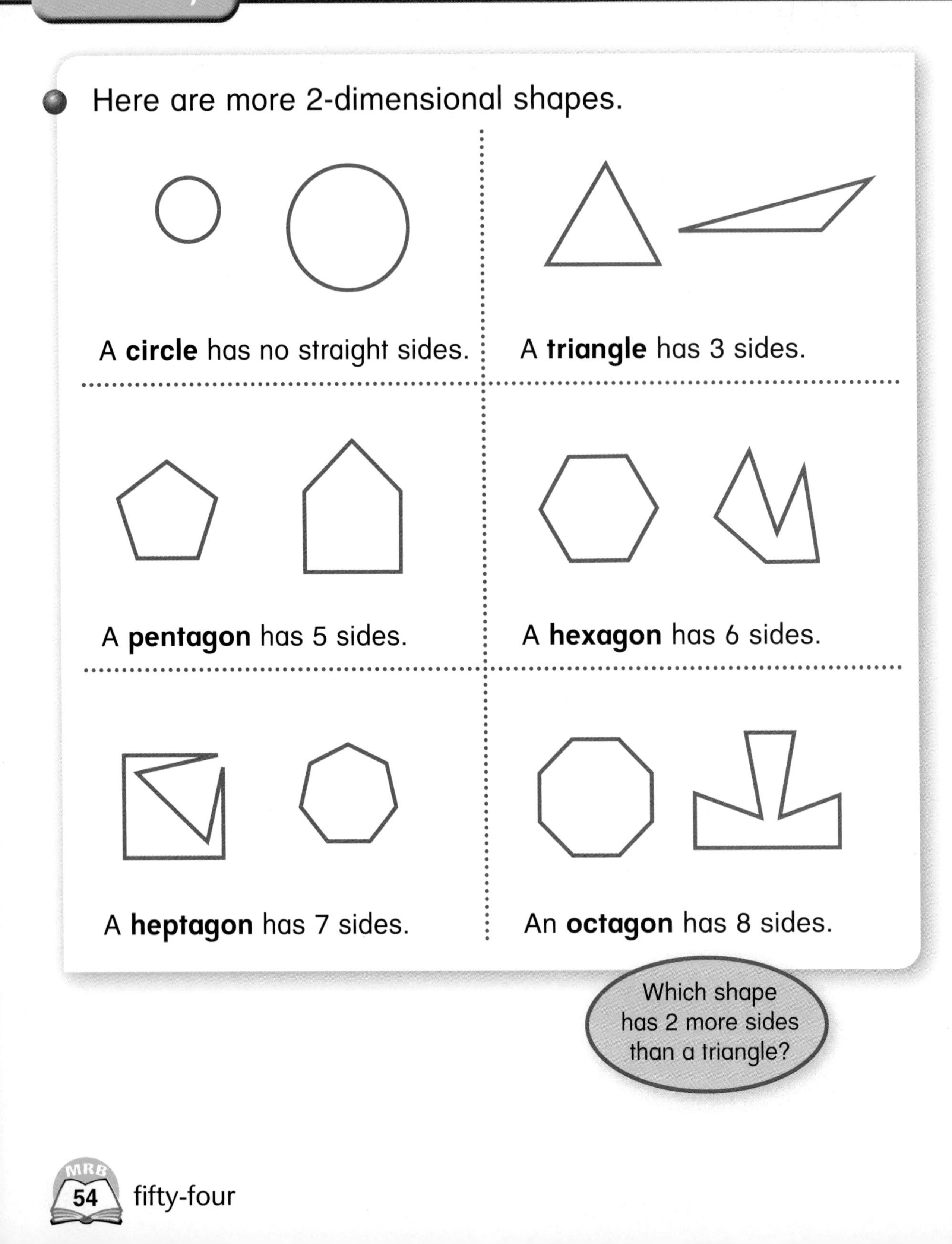
Here are more 2-dimensional shapes.
A **circle** has no straight sides.
A **triangle** has 3 sides.
A **pentagon** has 5 sides.
A **hexagon** has 6 sides.
A **heptagon** has 7 sides.
An **octagon** has 8 sides.
Which shape has 2 more sides than a triangle?

These 2-dimensional shapes are all **quadrangles.** Quadrangles are also called **quadrilaterals.**

- A quadrangle has 4 sides, 4 angles, and 4 vertices.

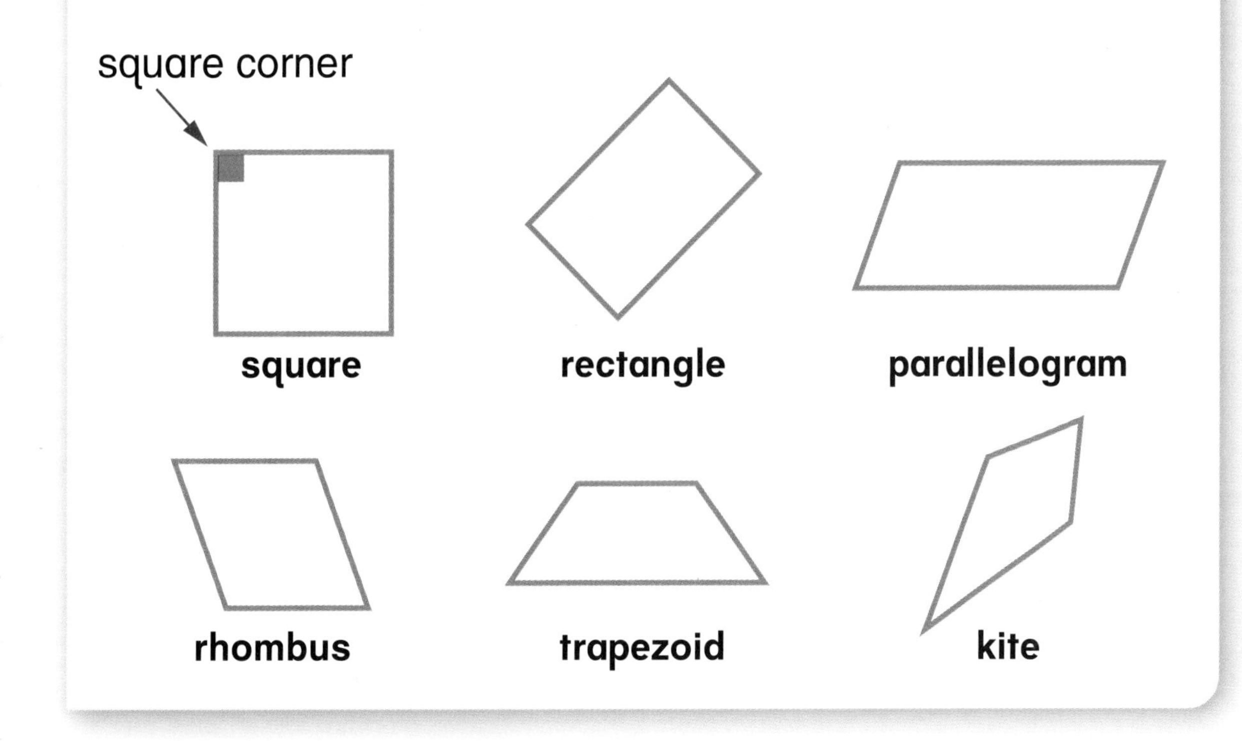

Try It Together

Choose one of the 2-dimensional shapes on pages 54–55. Describe it to a partner without saying the shape's name. Can your partner tell what shape you are describing?

3-Dimensional Shapes

Read It Together

A **3-dimensional** shape takes up space.
Objects you hold are 3-dimensional.

- Some 3-dimensional shapes are **rectangular prisms.**

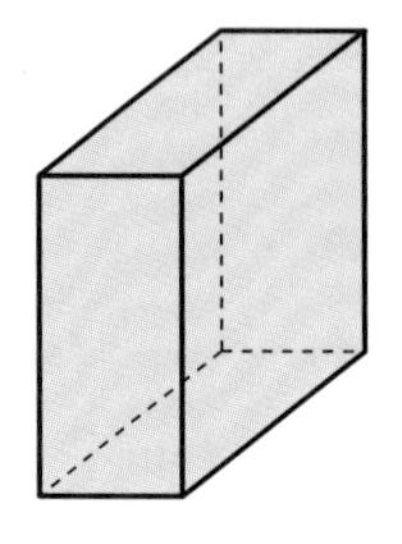

cube

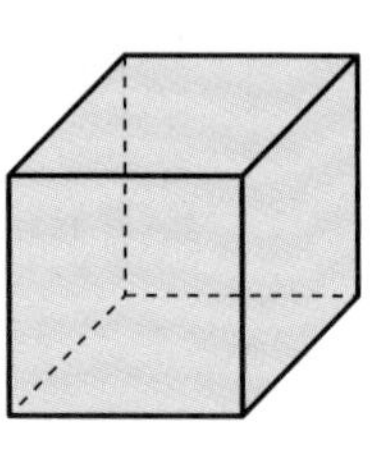

cube

- Some 3-dimensional shapes are **pyramids.**

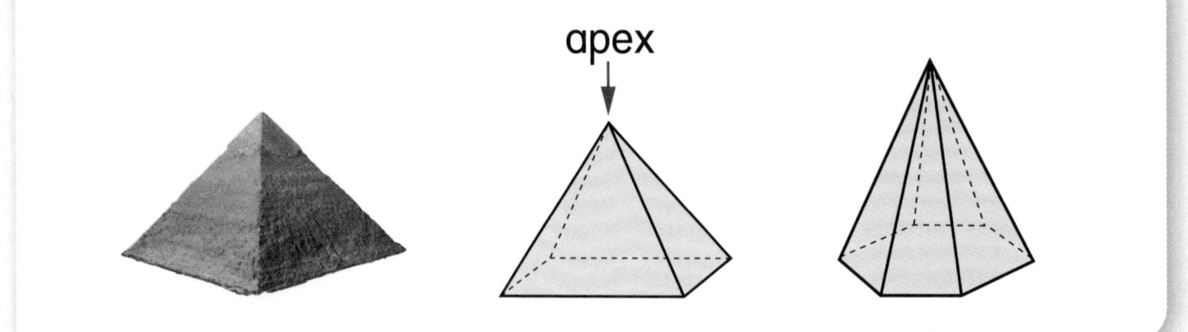

- Some 3-dimensional shapes are **cones.**
- Some 3-dimensional shapes are **cylinders.**
- Some 3-dimensional shapes are **spheres.**

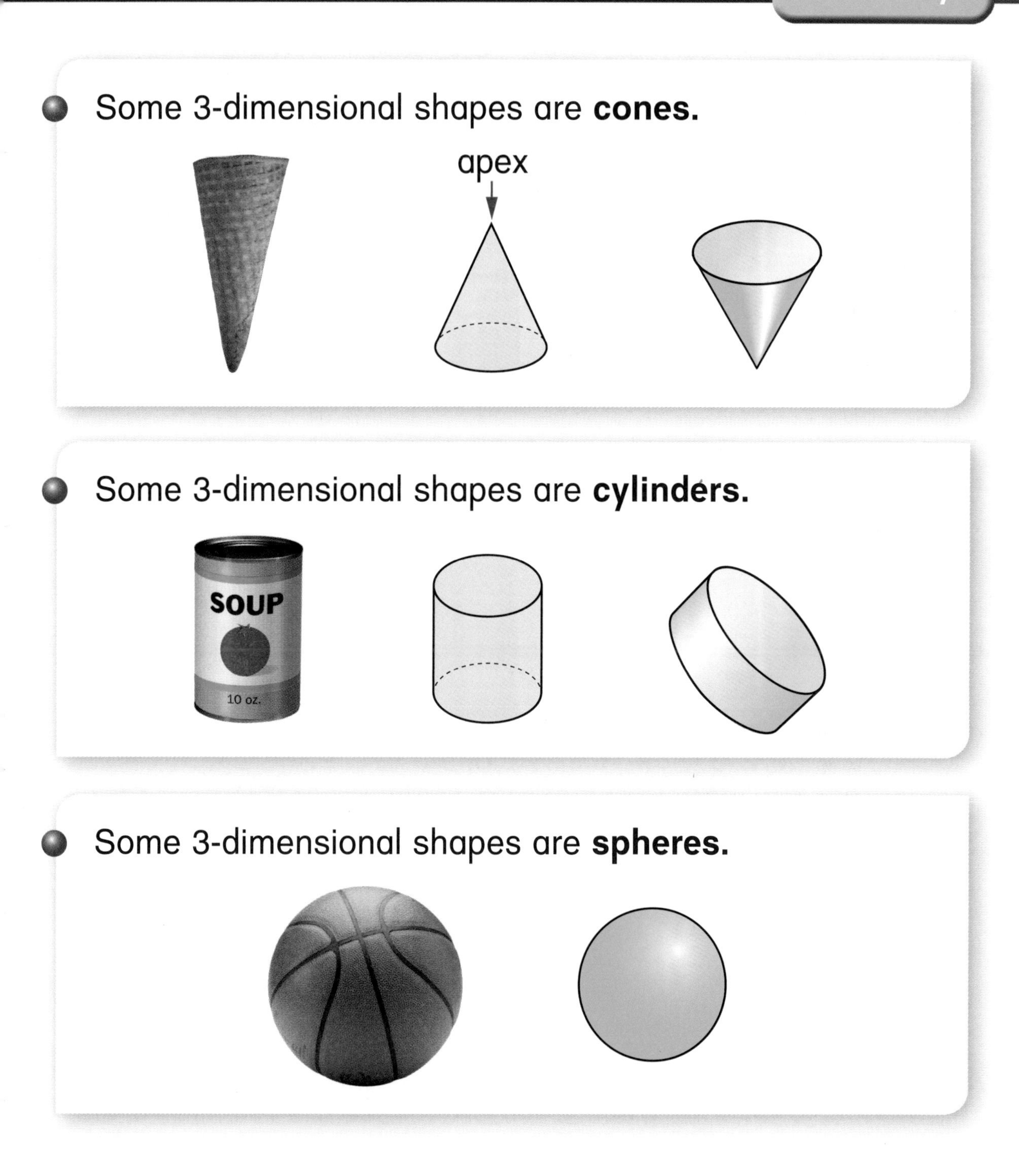

- Some 3-dimensional shapes have faces, edges, and vertices.

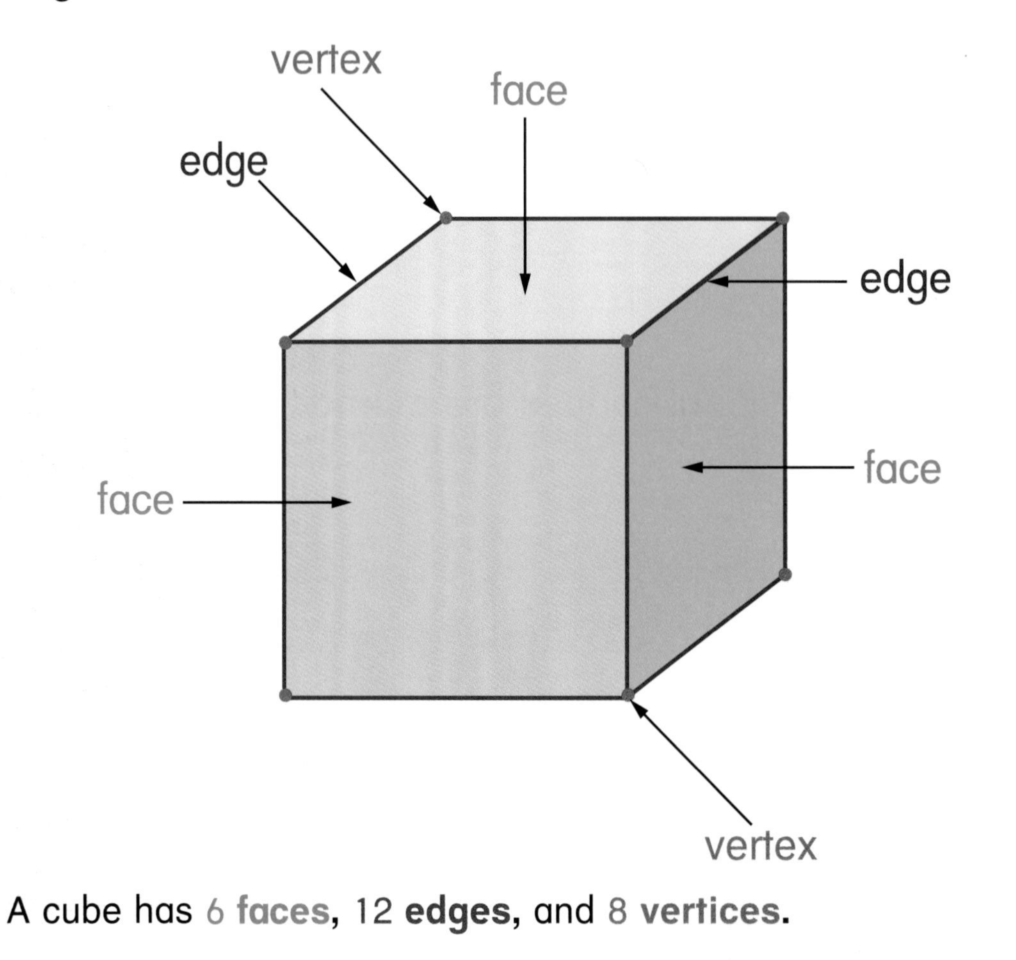

A cube has 6 **faces,** 12 **edges,** and 8 **vertices.**

Some of the faces on some 3-dimensional shapes are called **bases.**

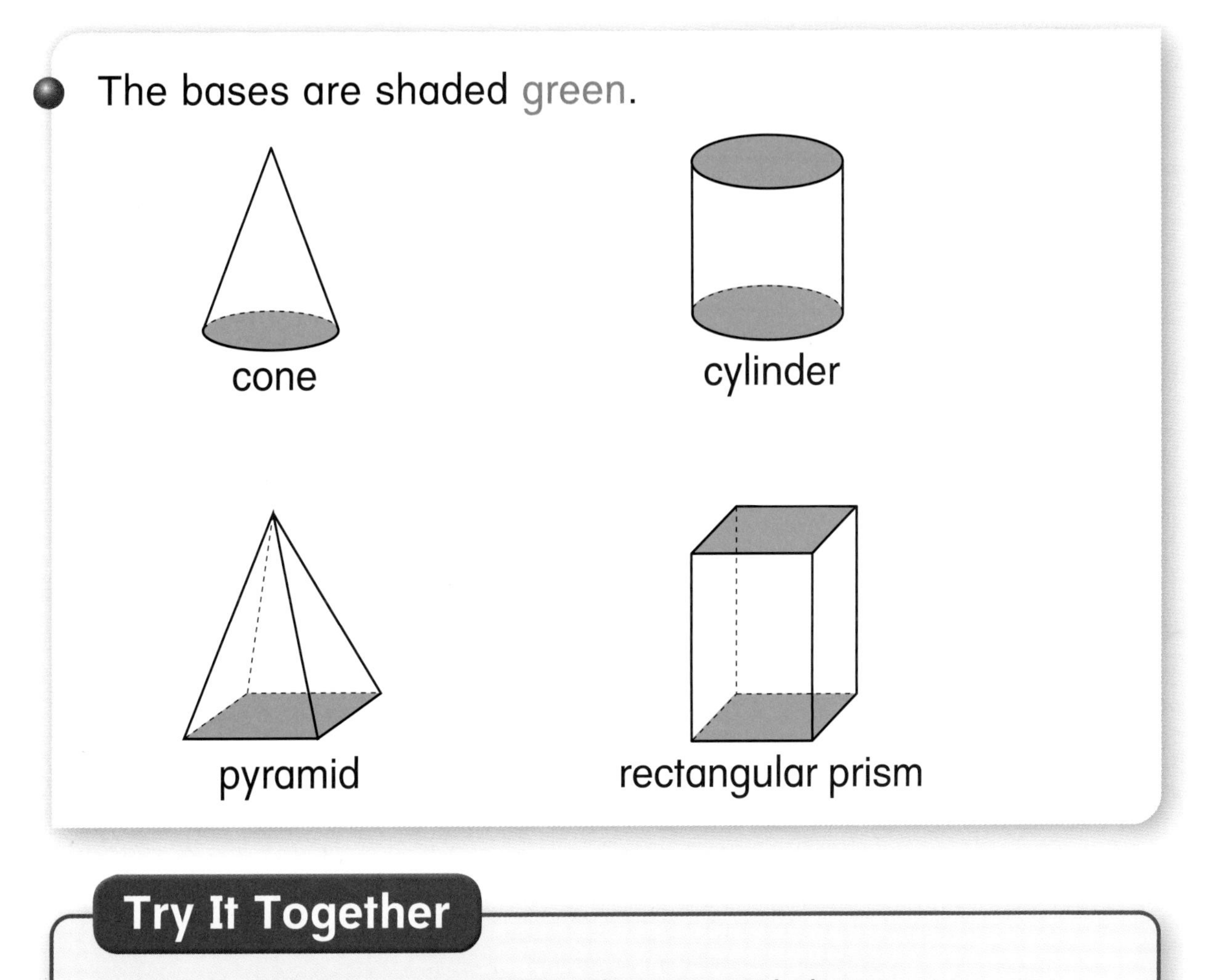

Try It Together

Look around the room. Find 3-dimensional shapes.
Can you find a rectangular prism? Can you find a sphere?

Line Symmetry

Read It Together

A shape has **line symmetry** if you can fold it in half and both halves match exactly. The fold line is called the **line of symmetry.**

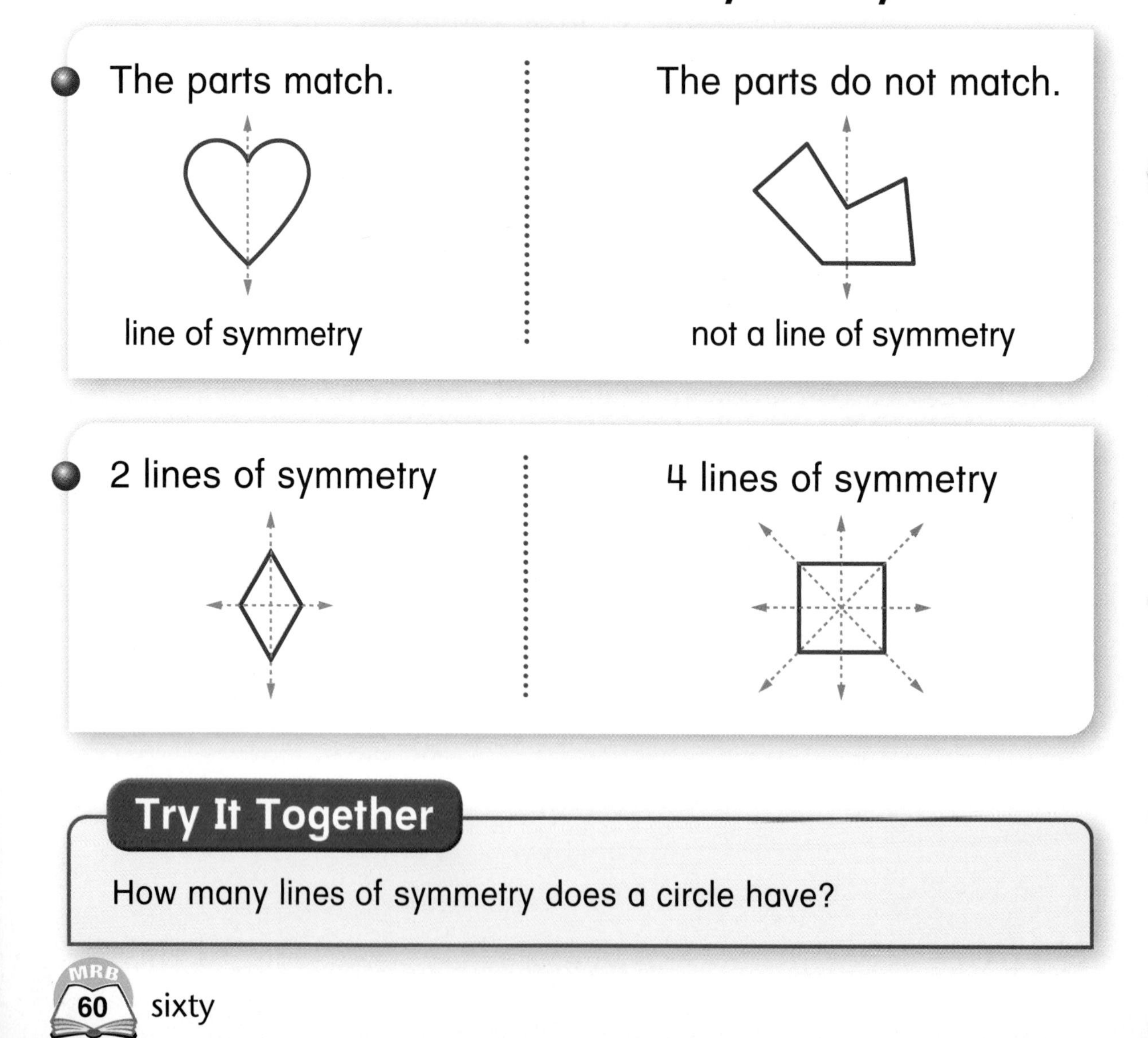

Try It Together

How many lines of symmetry does a circle have?

Measurement

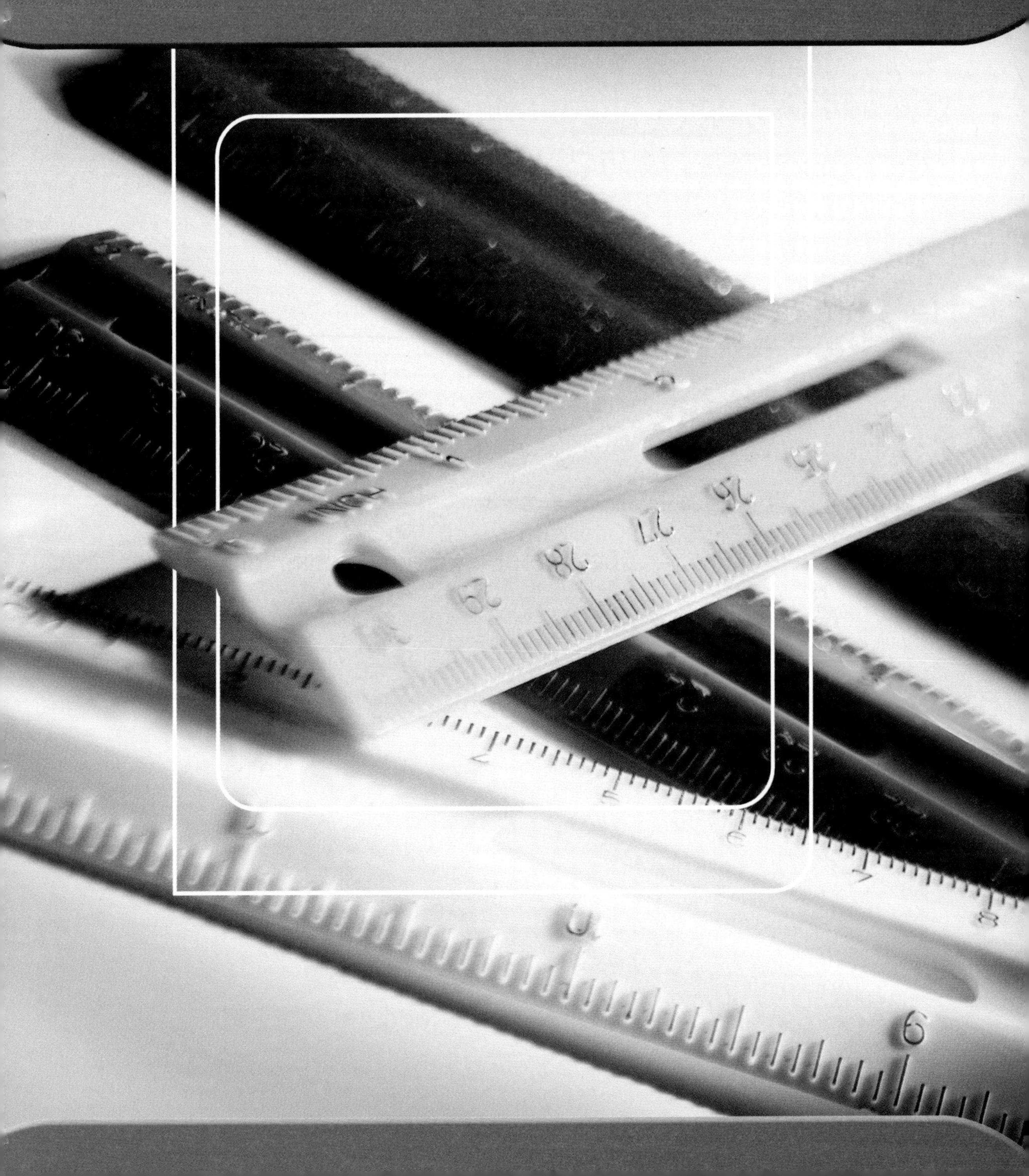

Length

Read It Together

You can use parts of your body to measure. Body measures tell us about how long something is. These measures are different for different people.

Here are some examples.

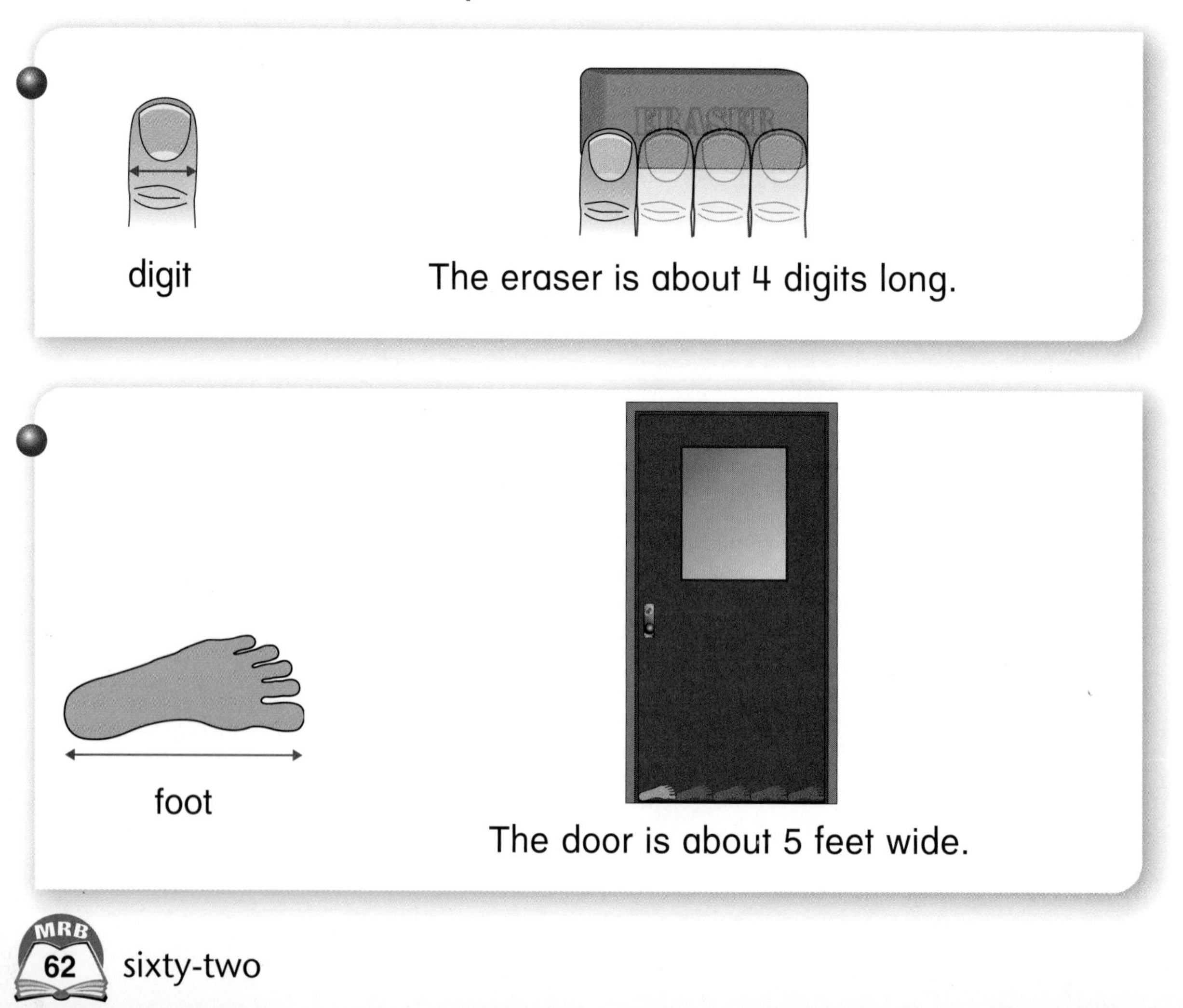

digit

The eraser is about 4 digits long.

foot

The door is about 5 feet wide.

arm span (or fathom)

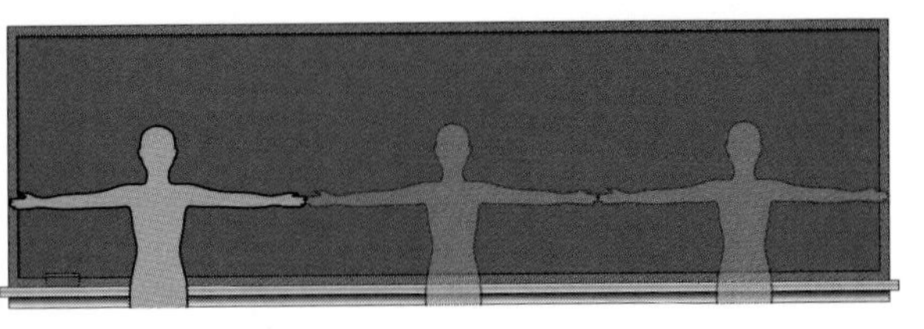

The chalkboard is about 3 arm spans wide.

Here are some more examples.

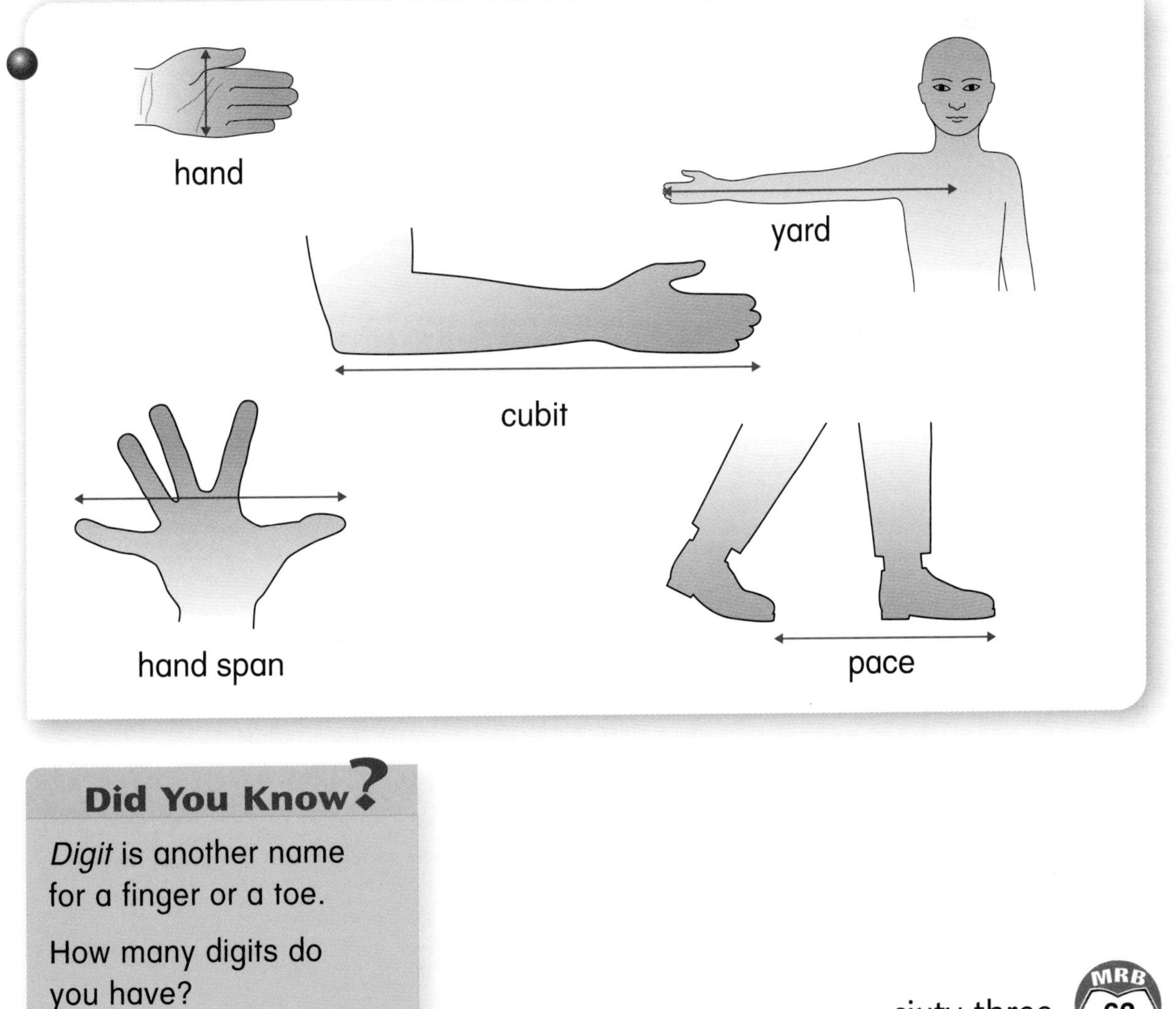

Did You Know?

Digit is another name for a finger or a toe.

How many digits do you have?

- You can use **tools** to measure length.
These tools use **standard units** that never change.
Standard units are the same for everyone.

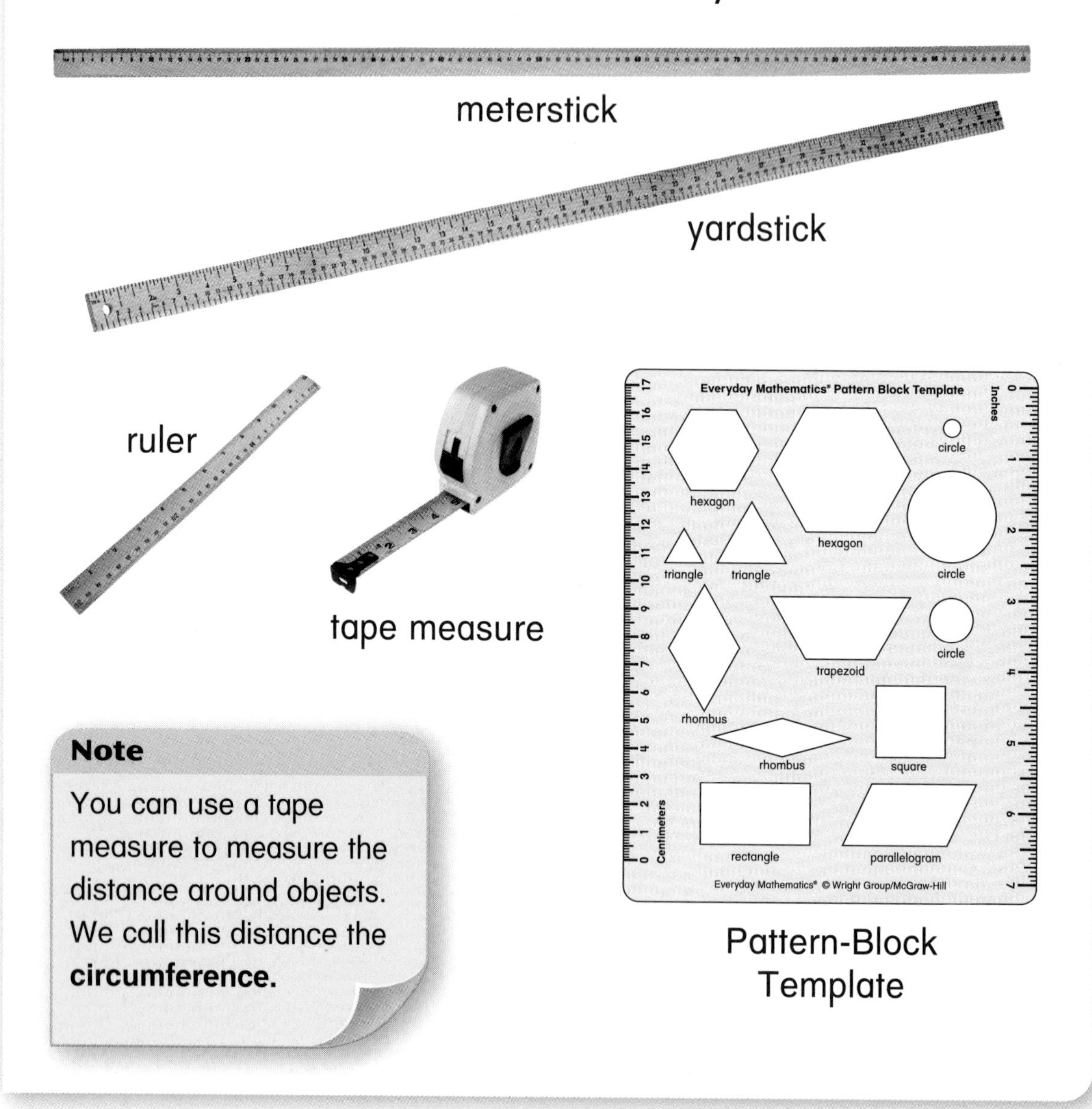

Note

You can use a tape measure to measure the distance around objects. We call this distance the **circumference.**

An **inch** is one standard unit.

Rulers are often marked with inches on one edge.

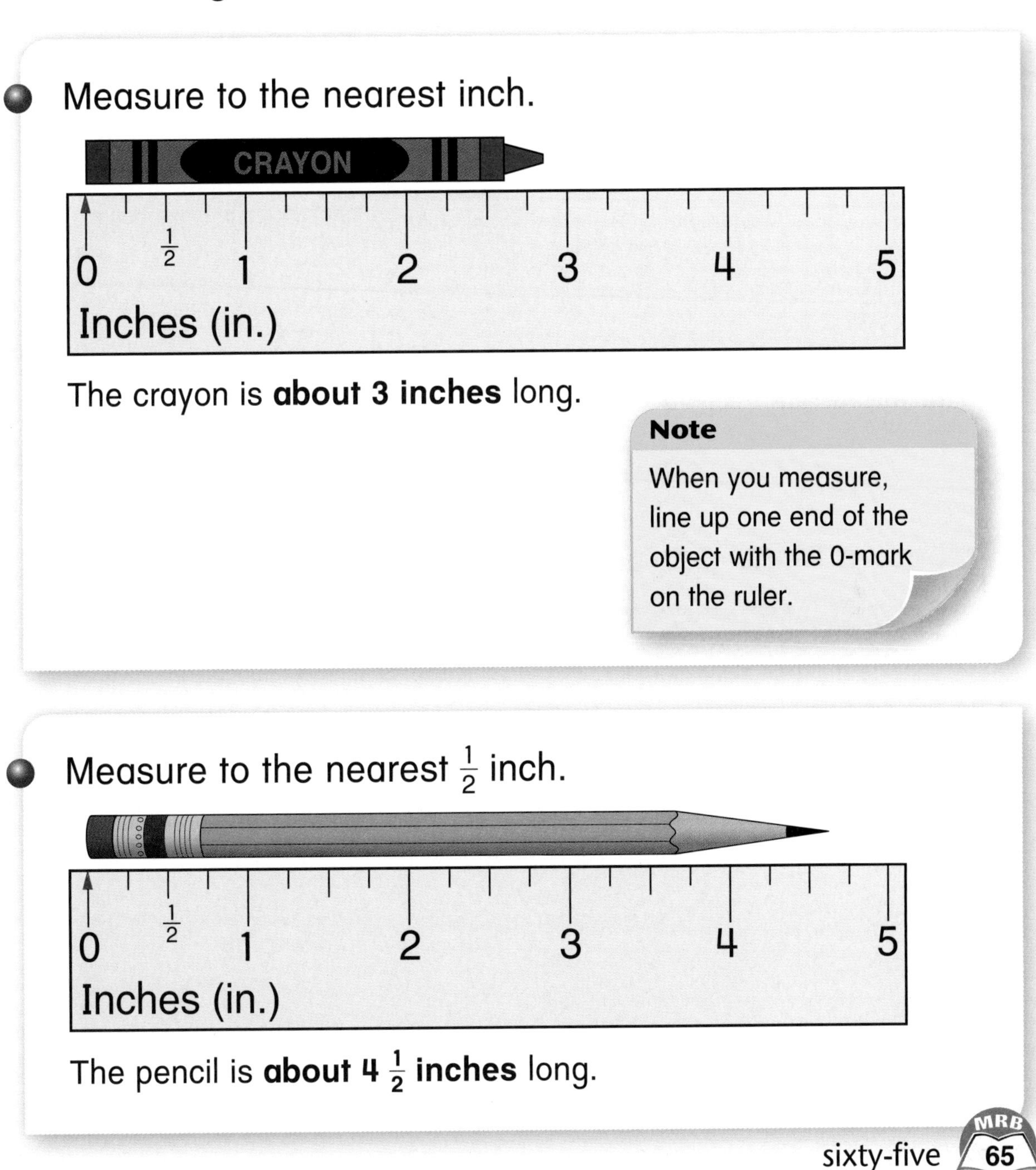

A **centimeter** is another standard unit.

Rulers are often marked with centimeters on one edge.

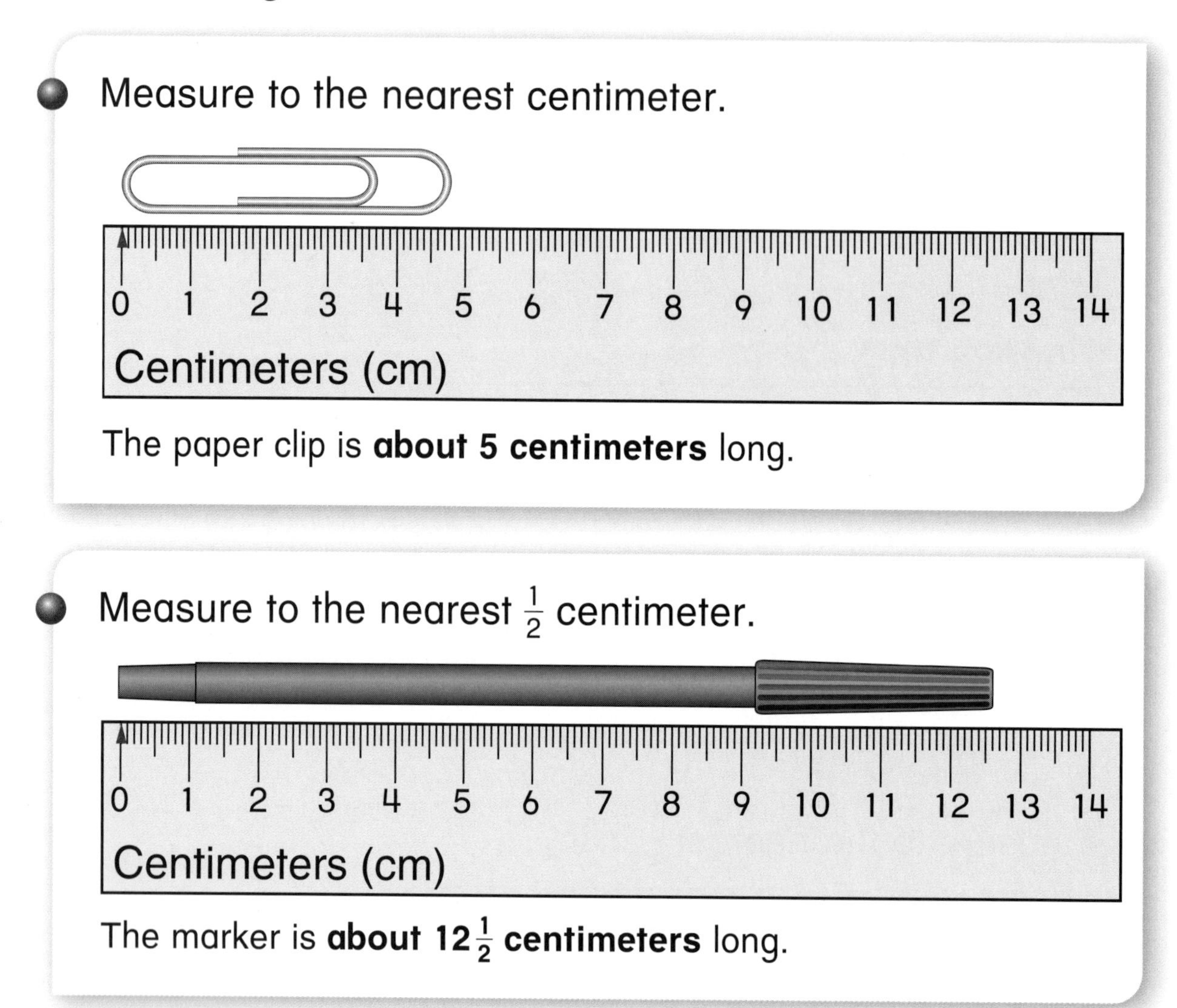

Today, people use two different measurement systems. One is the U.S. customary system, and the other is the metric system.

Here are some **units of length.**

Tables of Measures of Length

U.S. Customary Units
1 yard (yd) = 36 inches (in.)
= 3 feet (ft)
1 foot = 12 inches
= $\frac{1}{3}$ yard
1 inch = $\frac{1}{12}$ foot
= $\frac{1}{36}$ yard

Metric Units
1 meter (m) = 100 centimeters (cm)
= 10 decimeters (dm)
1 decimeter = 10 centimeters
= $\frac{1}{10}$ meter
1 centimeter = $\frac{1}{10}$ decimeter
= $\frac{1}{100}$ meter

Try It Together

Estimate the length of an object. Then measure it with a ruler.

Perimeter and Area

Read It Together

What are some ways you use measurement?

- Sometimes you want to know the distance around a shape. This distance is the **perimeter** of the shape.

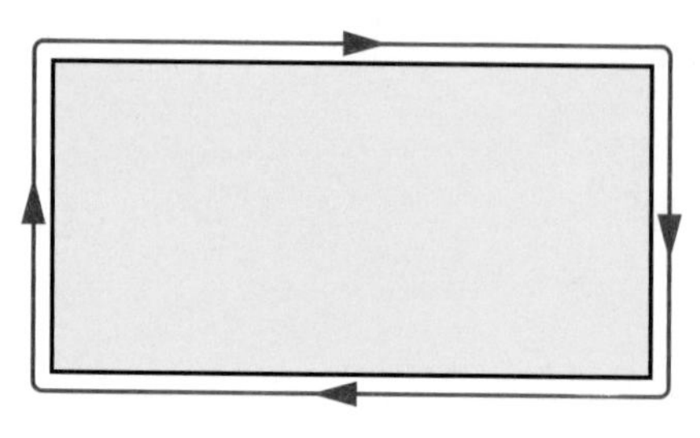

- Add the lengths of the sides to find the perimeter.

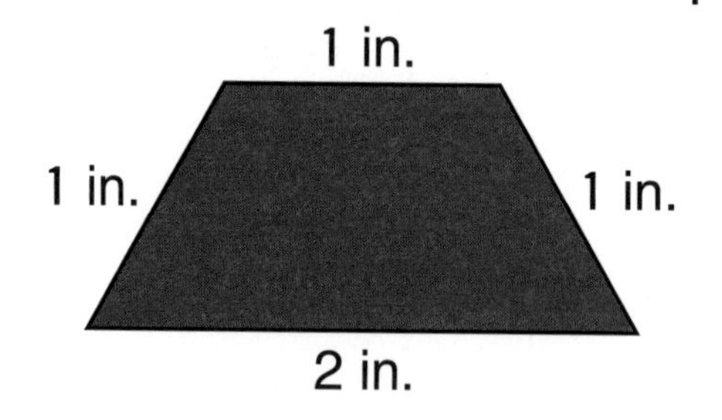

2 in. + 1 in. + 1 in. + 1 in. = 5 in.

The perimeter of the shape is 5 inches.

- Sometimes you want to know the amount of surface inside a shape. This amount of surface inside a shape is the **area** of the shape.

- Count the squares to find the area.

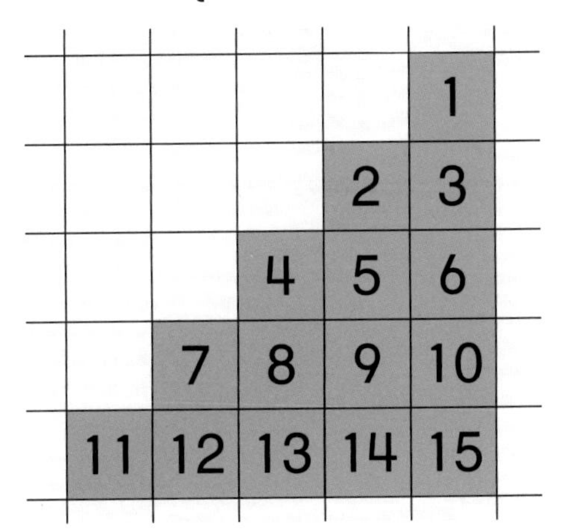

Note

Use *square* units for the area of a shape.

Each square is 1 square centimeter.

The area of the shape is 15 square centimeters.

Try It Together

Draw a shape that has a perimeter of 10 inches.
Compare your shape with a partner's shape.

Capacity and Weight

Read It Together

The amount a container can hold is called its **capacity.**

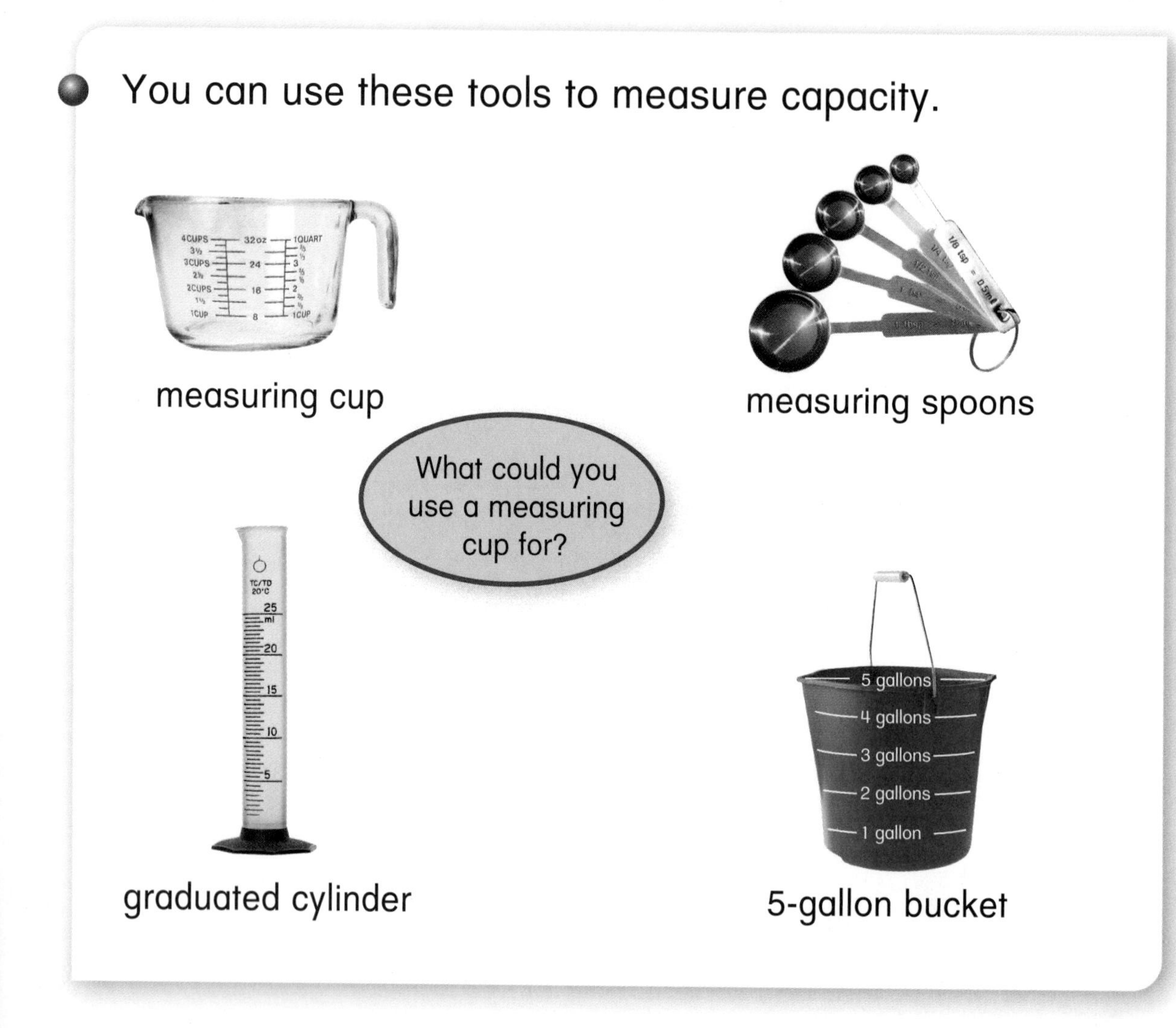

- You can use these tools to measure capacity.

measuring cup

measuring spoons

graduated cylinder

5-gallon bucket

Weight is the measure of how heavy something is.

- You can use these tools to measure weight.

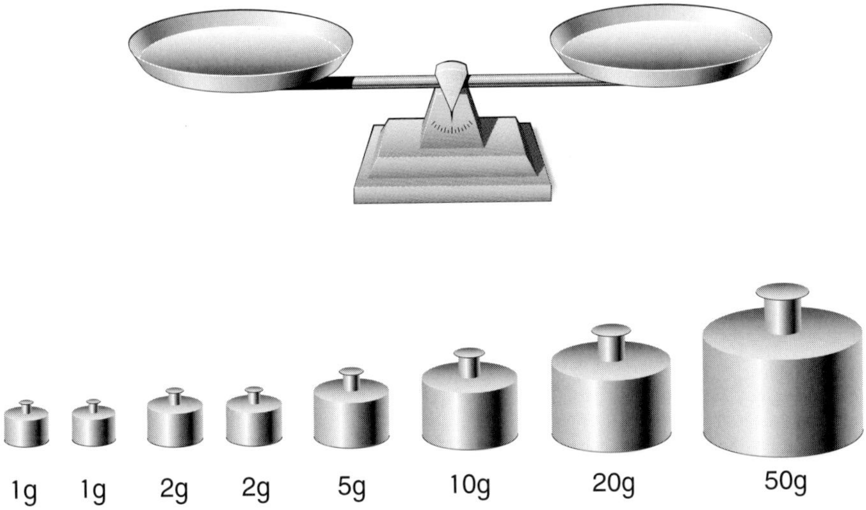

1g 1g 2g 2g 5g 10g 20g 50g

balance scale with weight set

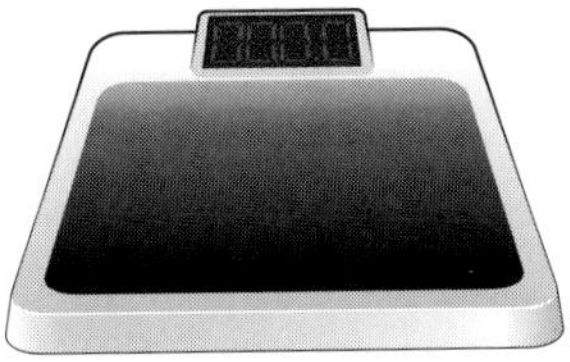

bath scale

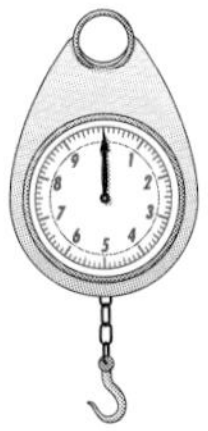

spring scale

Here are some **units of capacity.**

Tables of Measures of Capacity

U.S. Customary Units
1 cup (c) = $\frac{1}{2}$ pint
1 pint (pt) = 2 cups
1 quart (qt) = 2 pints
1 quart = 4 cups
1 gallon (gal) = 4 quarts
1 half-gallon ($\frac{1}{2}$ gal) = 2 quarts

Metric Units
1 liter (L) = 1,000 milliliters (mL)
$\frac{1}{2}$ liter = 500 milliliters

Here are some **units of weight.**

Tables of Measures of Weight

U.S. Customary Units
1 pound (lb) = 16 ounces (oz)
1 ton (T) = 2,000 pounds

Metric Units
1 kilogram (kg) = 1,000 grams (g)
1 metric ton (t) = 1,000 kilograms

Try It Together

How many cups are in a gallon?

Mathematics ... Every Day

Measures All Around: Animals and Tools

Have you ever wondered how large animals are measured? What about tiny animals or dangerous animals? People use many different tools to measure animals.

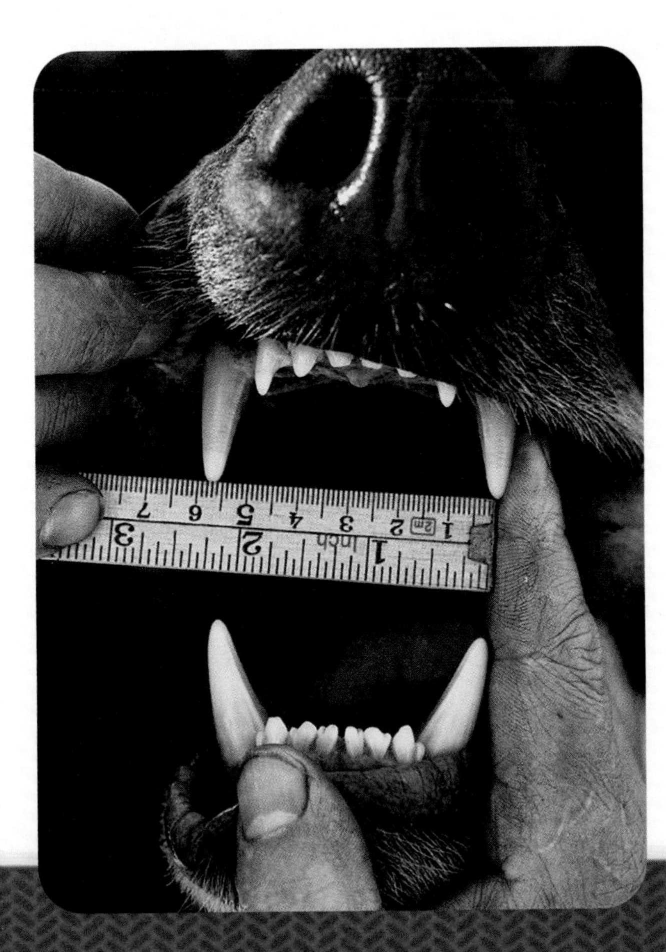

A bear researcher uses a ruler to measure the distance between a bear's canine teeth.

These scientists use calipers to measure the length of a sea turtle's shell and the length of a large egg. A caliper measures things from end to end. ➤

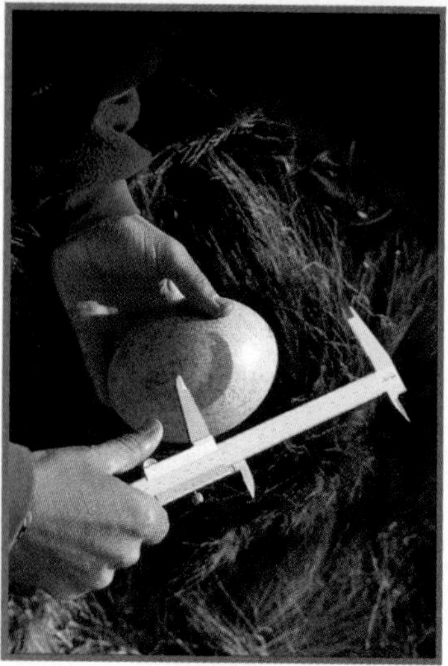

➤ A herpetologist studies reptiles and amphibians. This herpetologist uses a radio antenna to find the temperature of a snake. The antenna picks up signals from a tiny thermometer inside the snake.

Mathematics ... Every Day

An ornithologist studies birds. This ornithologist uses a spring scale to weigh an albatross chick.

This ornithologist uses a special thermometer to measure the surface temperature of a penguin.

These zookeepers use a large platform scale to weigh elephants. What other animals might need an extra-large scale?

A zookeeper uses a digital scale to weigh a cheetah. This scale measures weight in grams. Why might zookeepers want to know the weight of a young animal?

What tools do *you* use to measure?

Reference Frames

Time

Read It Together

There are many ways to show **time.** One way uses hour and minute hands and numbers to show time. Another way uses only numbers.

- You can use analog clocks and analog watches to show time.

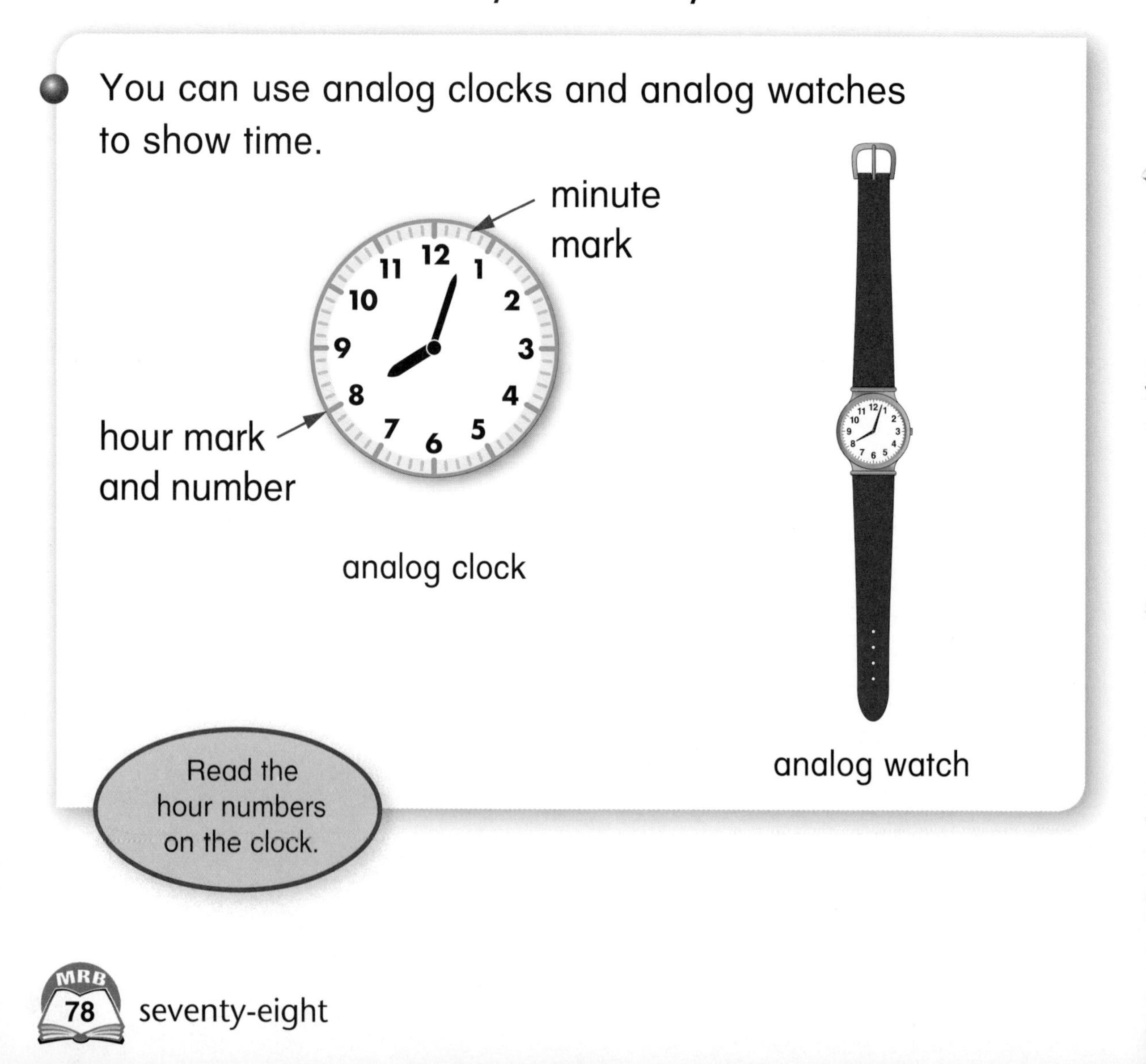

analog clock

analog watch

- You can use digital clocks and digital watches to show time.

digital clock

digital watch

Did You Know?

The first clock with a minute hand was invented in 1577.

The digital clock was invented in 1956.

The **hour hand** shows the hour.

The hour hand is usually the shorter hand on a clock face.

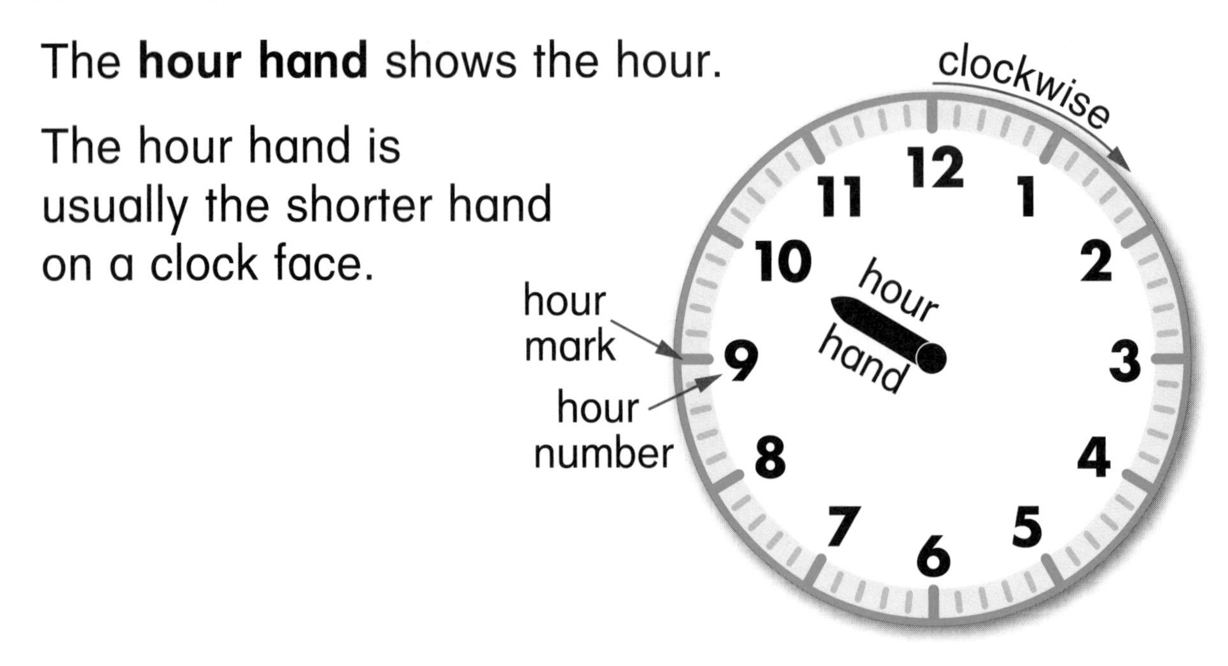

The hour hand takes 1 hour to move from an hour mark to the next hour mark.

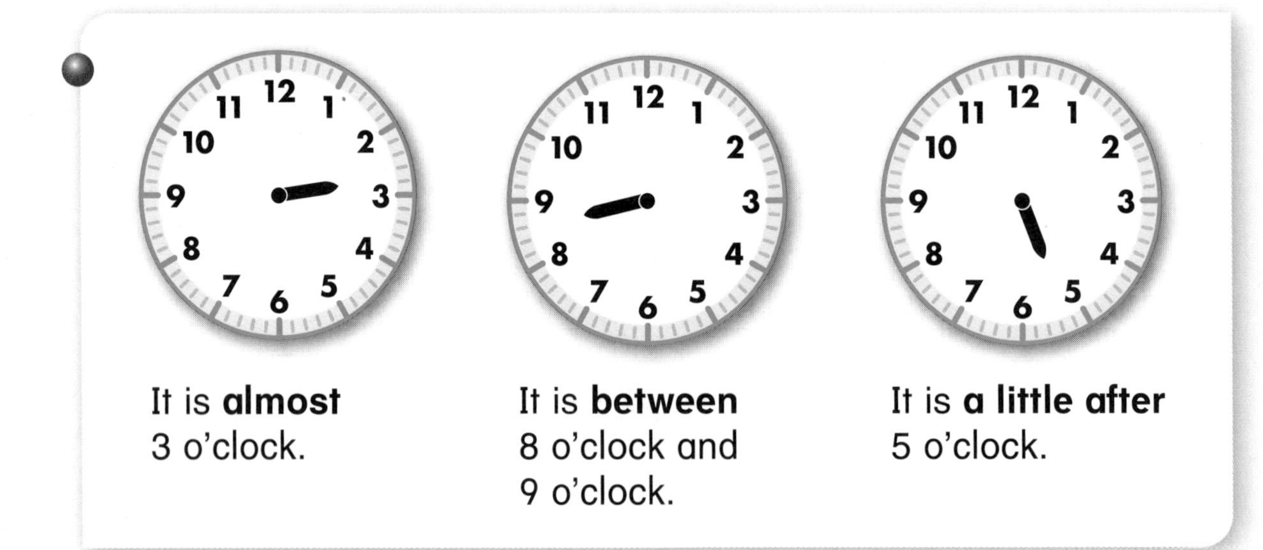

It is **almost** 3 o'clock.

It is **between** 8 o'clock and 9 o'clock.

It is **a little after** 5 o'clock.

The **minute hand** shows the minutes.

The minute hand is usually the longer hand on a clock face.

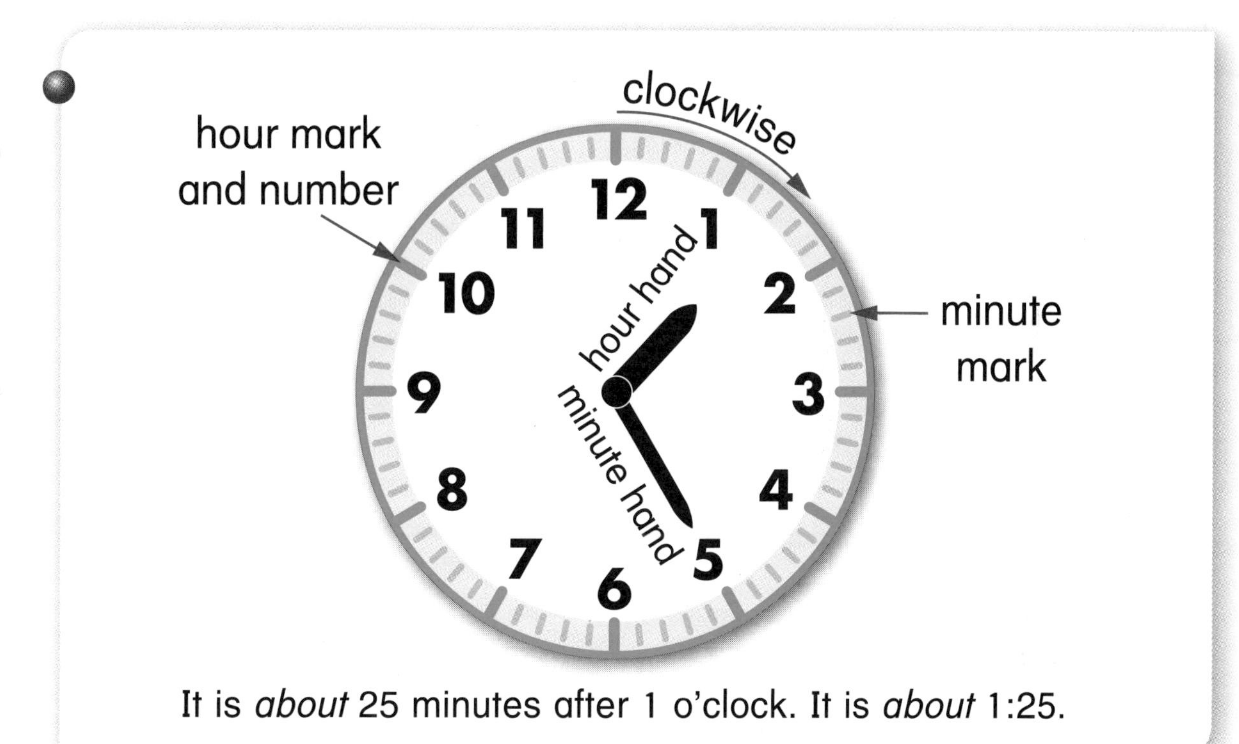

It is *about* 25 minutes after 1 o'clock. It is *about* 1:25.

The minute hand takes 1 minute to move from a minute mark to the next minute mark.

All hands on a clock move **clockwise.**

The hours from midnight to noon are **A.M.** hours.

12:00 A.M.
12 midnight

7:30 A.M.
half past 7

10:15 A.M.
a quarter after 10

12:00 P.M.
12 noon

The hours from noon to midnight are **P.M.** hours.

12:00 P.M.
12 noon

3:30 P.M.
half past 3

7:45 P.M.
a quarter to 8

12:00 A.M.
12 midnight

Calendars

Read It Together

We use a **calendar** to keep track of the days of the week and the months in a year.

Month   Year

March 2008						
Sun	Mon	Tues	Wed	Thurs	Fri	Sat
						1
2	3	4	5	6	7	8
9	10	11	12	13	14	15
16	17	18	19	20	21	22
23	24	25	26	27	28	29
30	31					

Abbreviation for days

Dates

March 25, 2008, is a Tuesday.

Month Date Year Day

What day of the week is March 6, 2008?

Every month has 30 or 31 days, except for one month.

Which month does not have 30 or 31 days?

Number of Days in Each Month

January	31 days	July	31 days
February	28 or 29* days	August	31 days
March	31 days	September	30 days
April	30 days	October	31 days
May	31 days	November	30 days
June	30 days	December	31 days

* 29 days in leap years

There is a **leap year** almost every four years.

We add an extra day to February in a leap year.

Did You Know?

It takes about $365\frac{1}{4}$ days for the Earth to make one complete trip around the sun. This is why we add a day almost every four years.

Here are some **units of time.**

Units of Time

Units of Time
1 minute = 60 seconds
1 hour = 60 minutes
1 day = 24 hours
1 week = 7 days
1 month = 28, 29, 30, or 31 days
1 year = 12 months
1 year = 52 weeks plus 1 day, or 52 weeks plus 2 days (in leap year)
1 year = 365 days, or 366 days (in leap year)
1 decade = 10 years
1 century = 100 years
1 millennium = 1,000 years

Try It Together

How many days are in 3 weeks? In 5 weeks?

Temperature

Read It Together

We use a **thermometer** to tell the **temperature.**

Temperature tells how hot or cold something is.

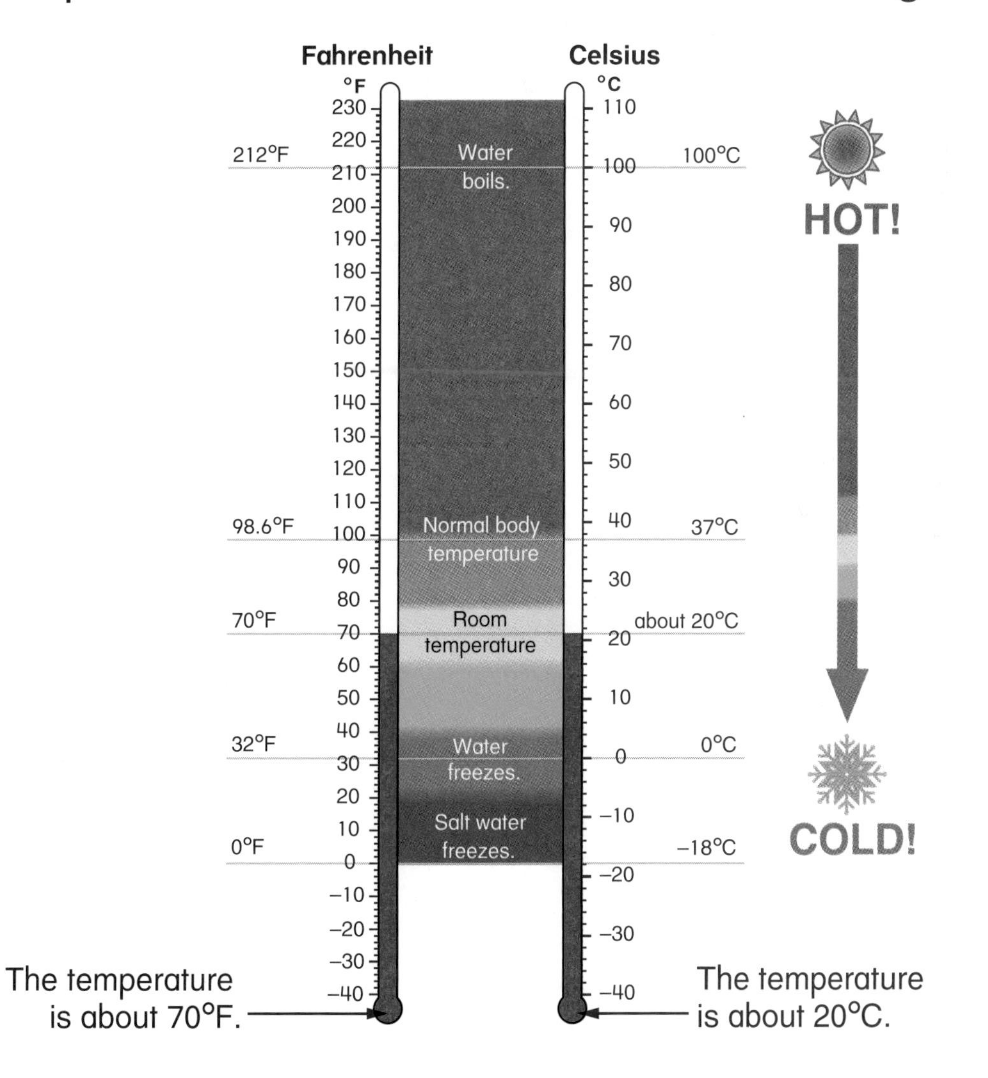

Money

Read It Together

We use **money** to buy things.

Here are some of the coins and bills used in the United States.

	Penny 1¢ $0.01	Nickel 5¢ $0.05
Heads or Front		
Tails or Back		
Equivalencies	1 Ⓟ	1 Ⓝ 5 Ⓟ

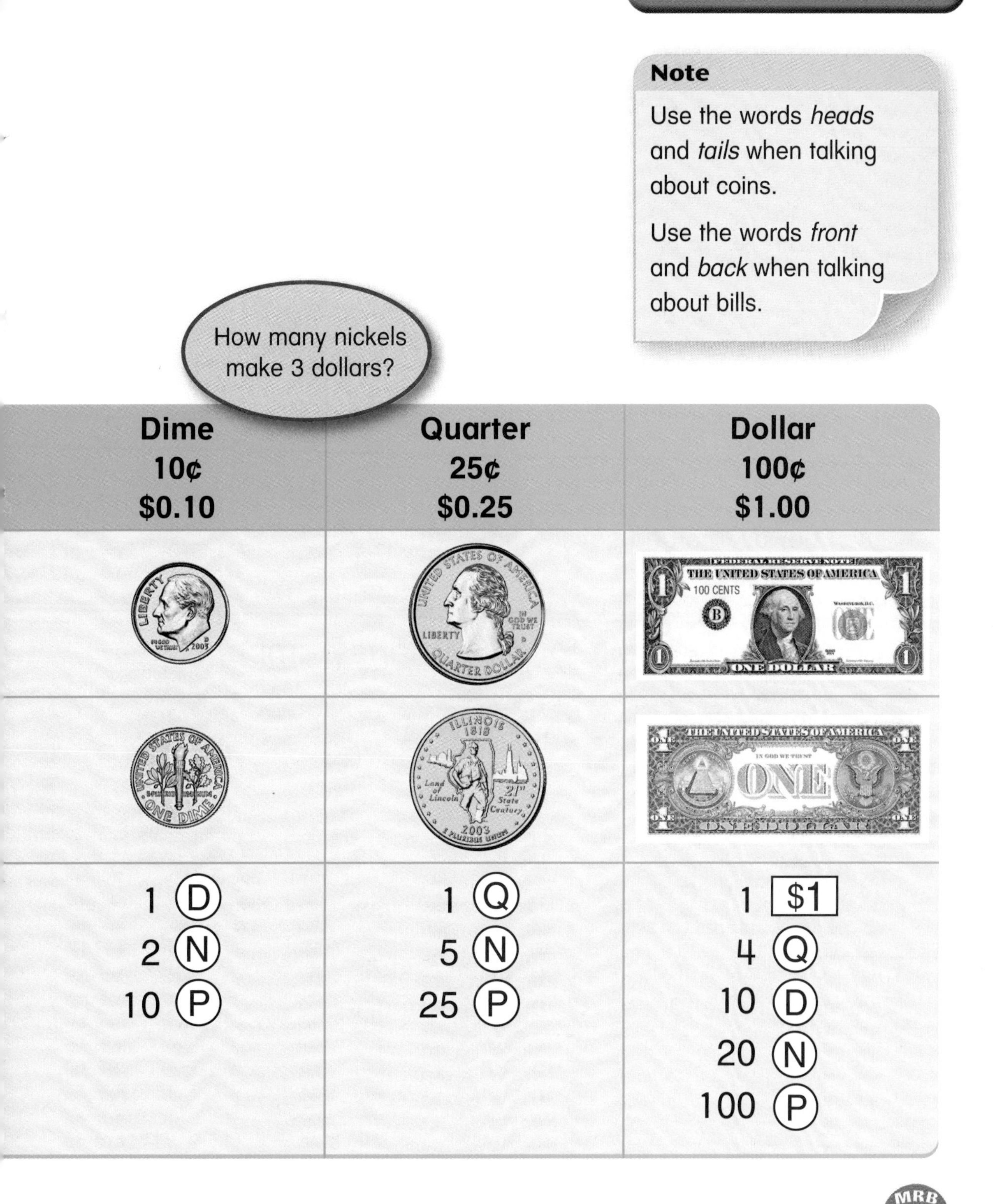

Note

Use the words *heads* and *tails* when talking about coins.

Use the words *front* and *back* when talking about bills.

Dime **10¢** **\$0.10**	**Quarter** **25¢** **\$0.25**	**Dollar** **100¢** **\$1.00**
1 (D) 2 (N) 10 (P)	1 (Q) 5 (N) 25 (P)	1 [\$1] 4 (Q) 10 (D) 20 (N) 100 (P)

You can use **dollars-and-cents notation** to write money amounts.

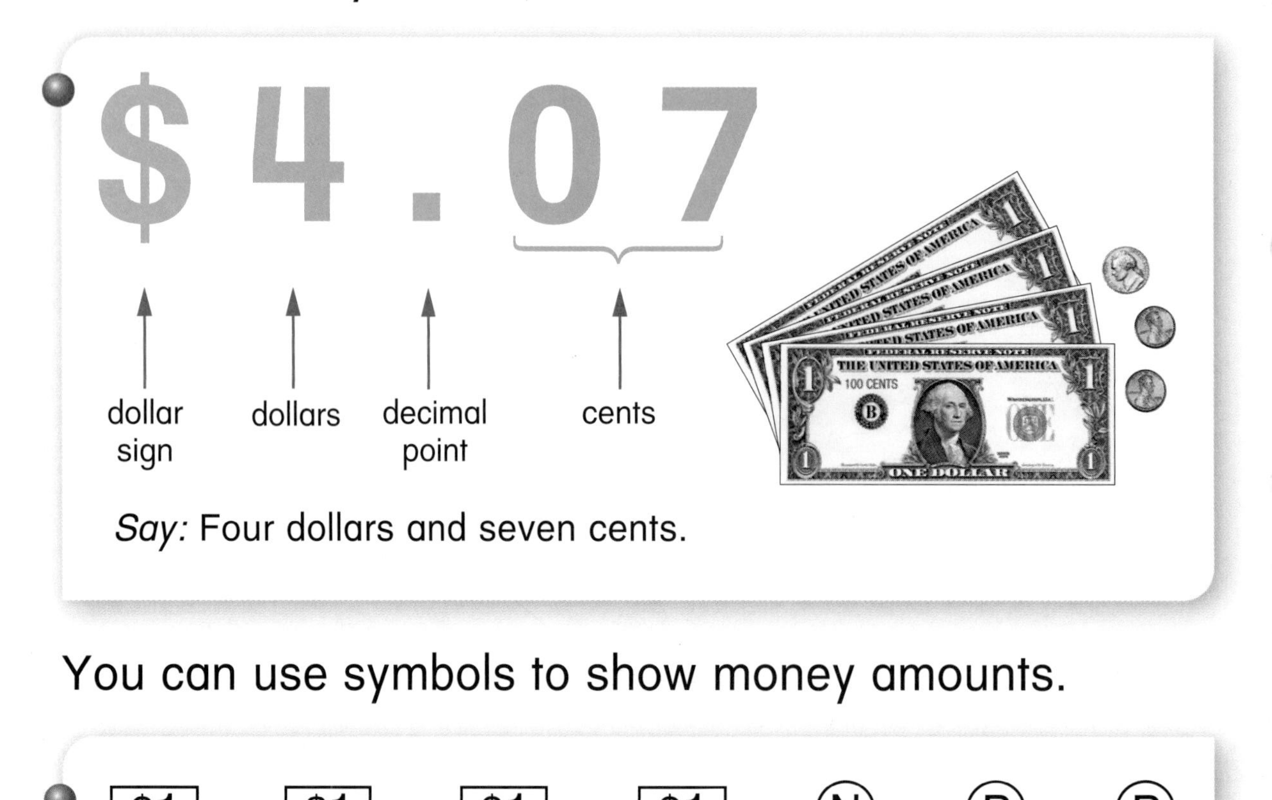

Say: Four dollars and seven cents.

You can use symbols to show money amounts.

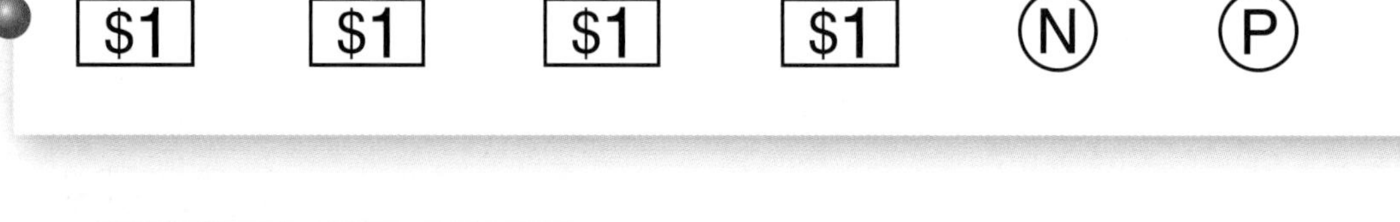

Try It Together

Play "Coin Exchange" on pages 128 and 129 or "One-Dollar Exchange" on pages 144 and 145.

Estimation

Estimation

Read It Together

An **estimate** is an answer that should be close to an exact answer. Sometimes an estimate is called a **ballpark estimate.** We can use ballpark estimates—estimates that are "close enough"—to check answers.

You make estimates every day. When you say the word *about* before a number, you are making an estimate.

One way to estimate a large number of objects is to look at a smaller part.

- About how many cubes are in the jar?

There are about 25 cubes on the top layer and 6 layers of cubes in the jar.
Four 25s is 100
and two 25s is 50.
100 plus 50 equals 150.

There are about 150 cubes in the jar.

Another way to find an estimate is to change numbers to **close-but-easier** numbers.

- There are 100 cartons of milk in the lunchroom.
  53 first graders need milk.
  38 second graders need milk.
  Is there enough milk in the lunchroom for the first and second graders?

Close-but-easier numbers

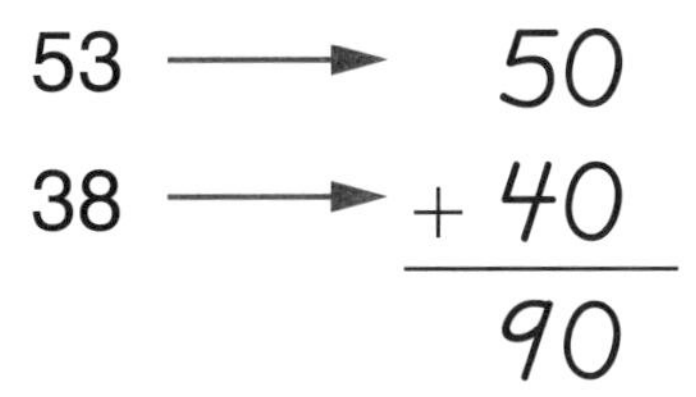

About 90 cartons of milk are needed. So 100 cartons is enough for all of the first and second graders.

What word lets someone know you are making an estimate?

Use an estimate to check if an answer makes sense.

- Anna's class checked out 94 library books last month.
  37 books were returned.
  How many books are still checked out?

  Add up to find the answer.

  +3 +50 +4

  37 → 40 → 90 → 94

  3 + 50 + 4 = 57

  57 books are still checked out.

  Estimate to check the answer.

  Close-but-easier numbers

  94 → 90

  37 → − 40

  50

  Since 50 is close to 57, the answer makes sense.

Patterns and Functions

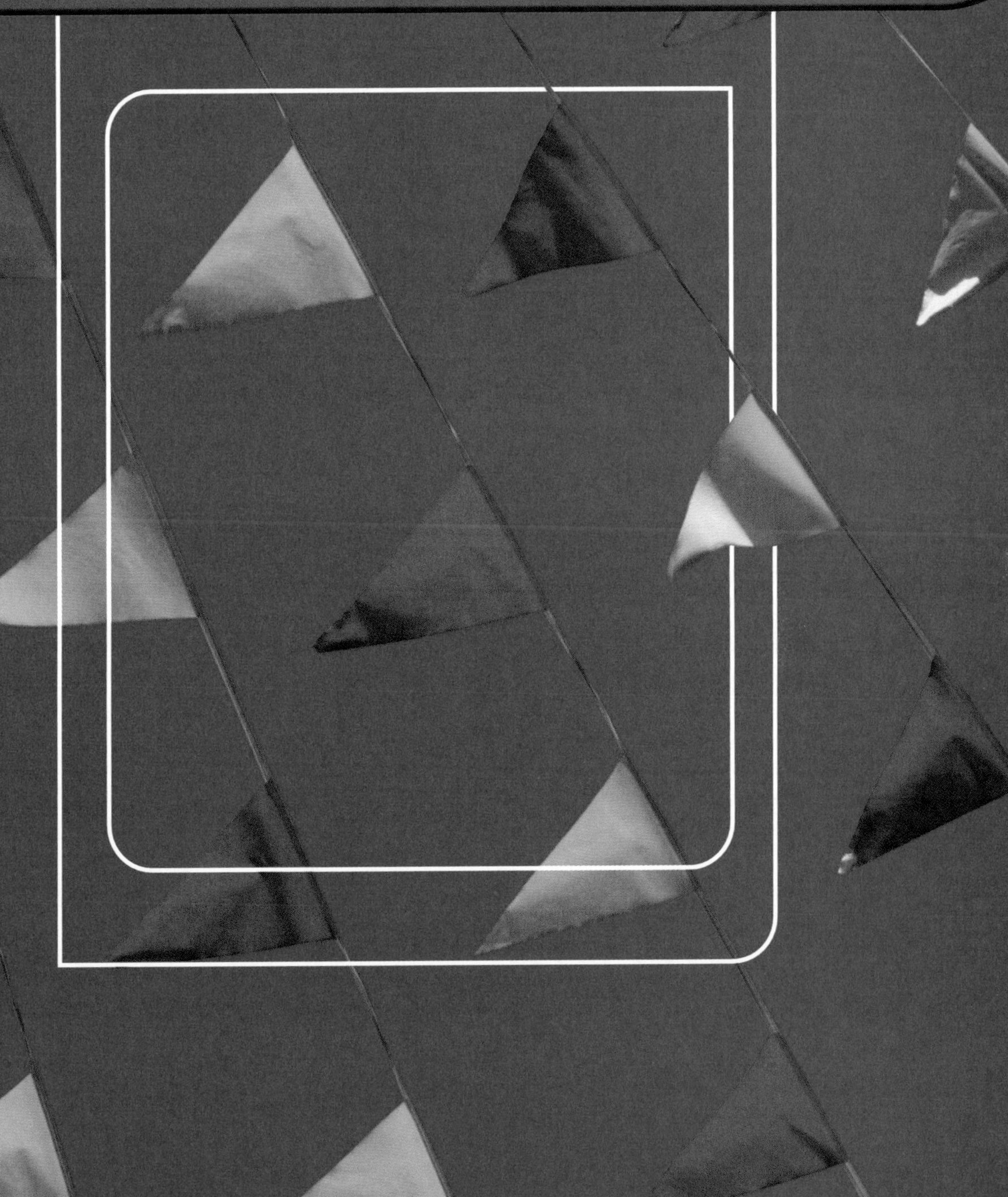

Patterns

Read It Together

Shapes can make **patterns.** You can tell what comes next in a pattern if you know the rule.

- Some patterns repeat over and over.

Numbers can also make patterns.

- Even numbers and odd numbers can make dot patterns.

Even Numbers

2 4 6 8

Odd Numbers

1 3 5 7

A number grid has many patterns.

- This number grid shows odd and even numbers.

−9	−8	−7	−6	−5	−4	−3	−2	−1	0
1	2	3	4	5	6	7	8	9	10
11	12	13	14	15	16	17	18	19	20
21	22	23	24	25	26	27	28	29	30
31	32	33	34	35	36	37	38	39	40
41	42	43	44	45	46	47	48	49	50
51	52	53	54	55	56	57	58	59	60
61	62	63	64	65	66	67	68	69	70
71	72	73	74	75	76	77	78	79	80
81	82	83	84	85	86	87	88	89	90
91	92	93	94	95	96	97	98	99	100
101	102	103	104	105	106	107	108	109	110

The **odd** numbers are green.
The **even** numbers are orange.

Try It Together

Look at these numbers. 203, 205, 207, 209, …
Tell a partner the number that comes next.

Frames and Arrows

Read It Together

In a **Frames-and-Arrows diagram,** the **frames** are the shapes that hold the numbers, and the **arrows** show the path from one frame to the next.

Each diagram has a **rule box.** The **rule** in the box tells how to get from one frame to the next.

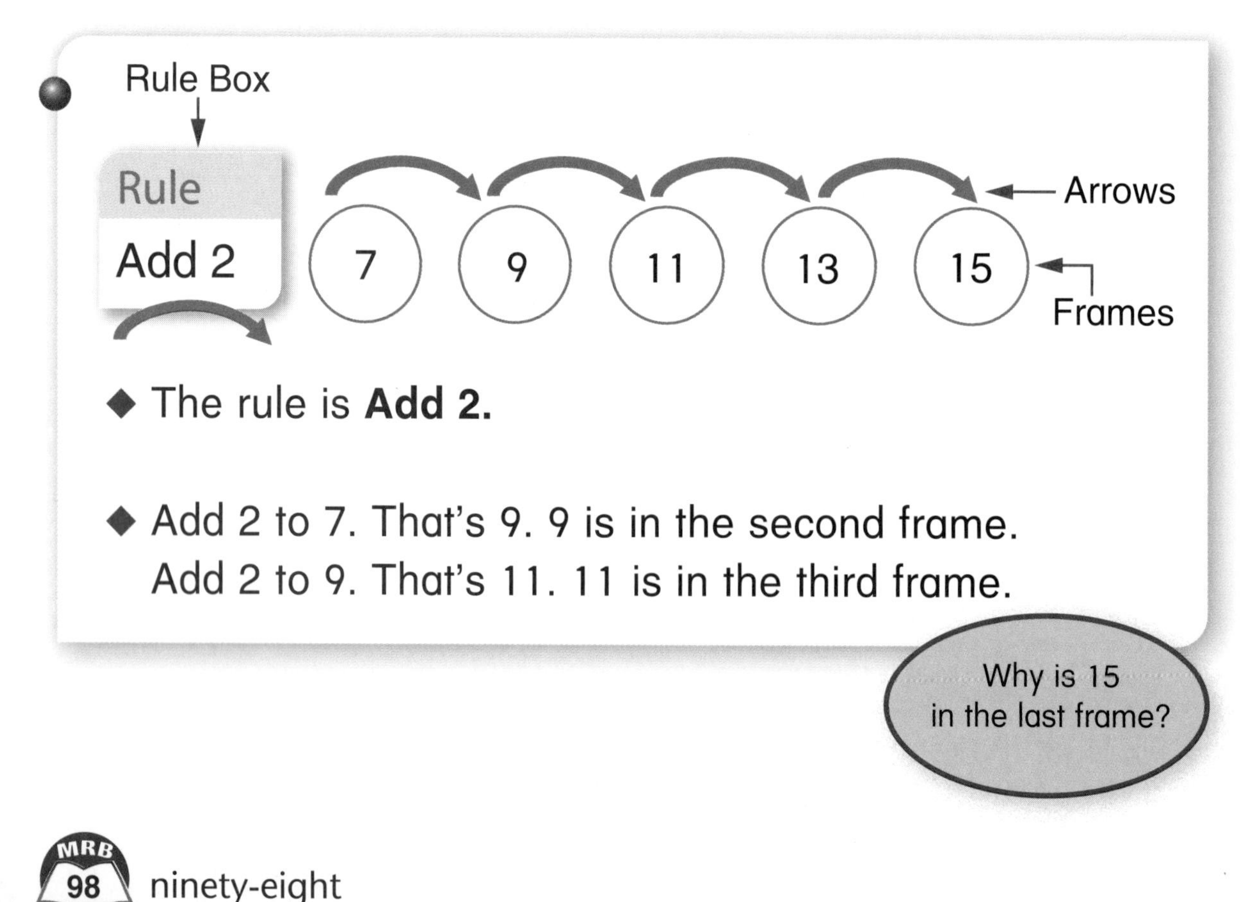

- The rule is **Add 2.**
- Add 2 to 7. That's 9. 9 is in the second frame.
  Add 2 to 9. That's 11. 11 is in the third frame.

Why is 15 in the last frame?

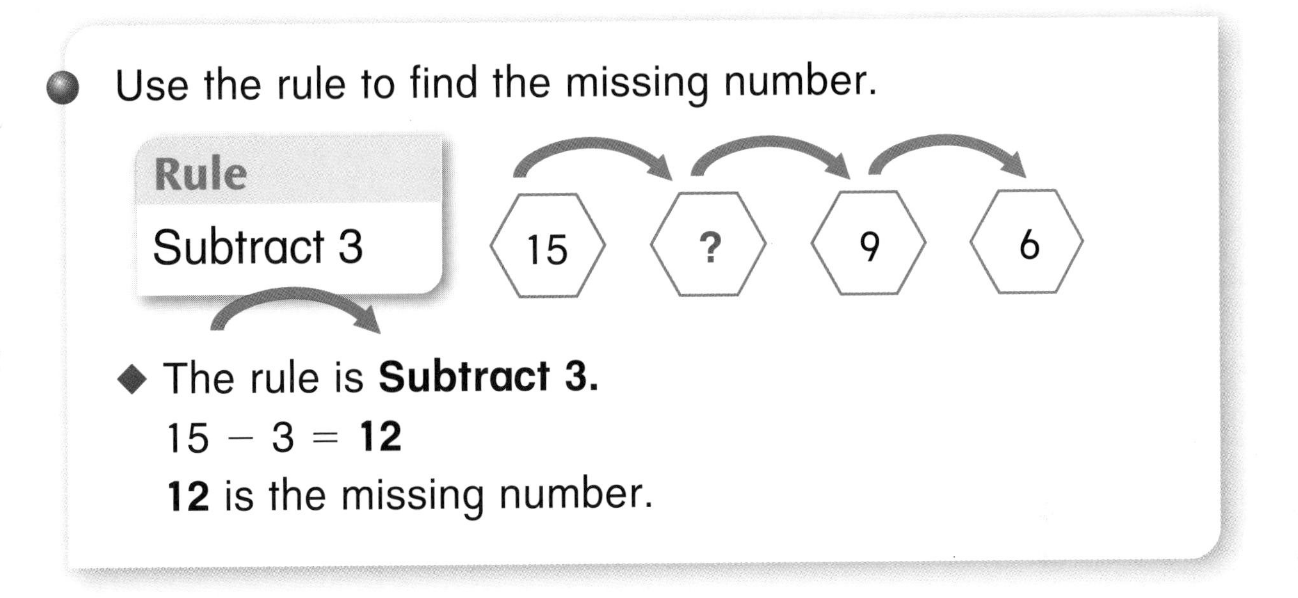

◆ The rule is **Subtract 3.**

15 − 3 = **12**

12 is the missing number.

Use the numbers in the frames to find the rule.

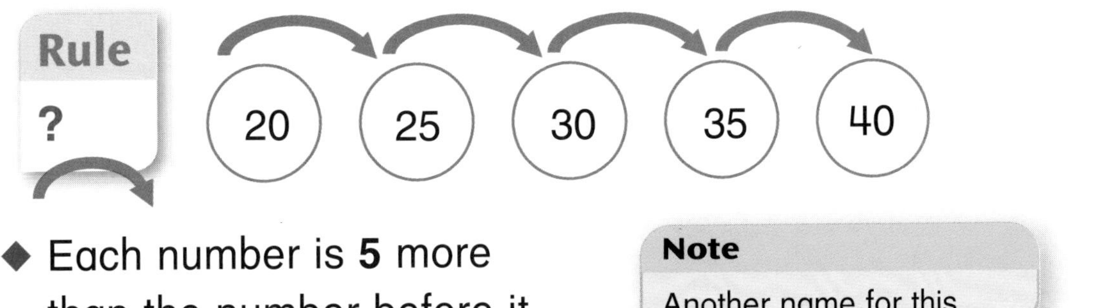

◆ Each number is **5** more than the number before it. The rule is **Add 5.**

Note

Another name for this rule is **Count up 5.**

Function Machines

Read It Together

A **function machine** uses a rule to change numbers. You put a number into the machine. The machine uses the rule to change the number. The changed number comes out of the machine.

- If you put 2 into the machine, it will **add 10** to 2. The number 12 will come out.
- If you put 4 into the machine, it will **add 10** to 4. The number 14 will come out.
- If you put 0 into the machine, it will **add 10** to 0. The number 10 will come out.

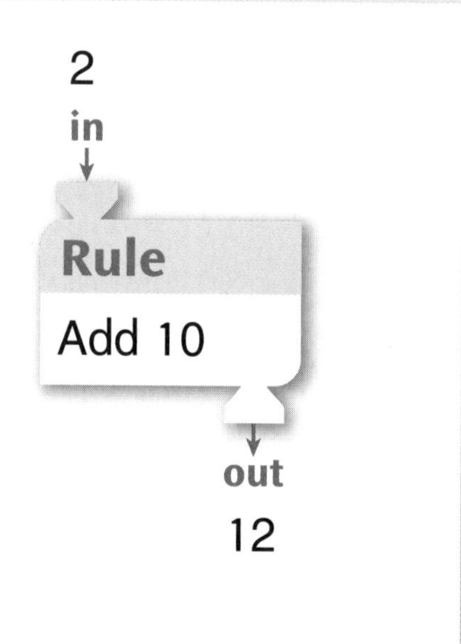

An **In and Out** table keeps track of how a function machine changes numbers.

in
↓
Rule
Subtract 3
↓
out

in	out
4	1
5	2
6	3
7	4

Write the numbers that are put into the machine in the **in** column. Write the numbers that come out of the machine in the **out** column.

- If you put 4 into the machine, the machine subtracts 3 from 4. 1 comes out.
- If 2 comes out, then 5 was put in. The machine subtracted 3 from 5.

- Use the in and out numbers to find a rule for a function machine.

in	out
2	0
5	3
6	4
9	7

◆ How do you get from 2 to 0? **Subtract 2.**
How do you get from 5 to 3? **Subtract 2.**
Can you **subtract 2** to get from 6 to 4? Yes.

◆ The rule is **Subtract 2.**

Try It Together

Ask your partner to write a rule. Write an In and Out table for the rule.

Patterns All Around

Patterns are everywhere! You can see patterns in nature. You can see patterns in the things people make.

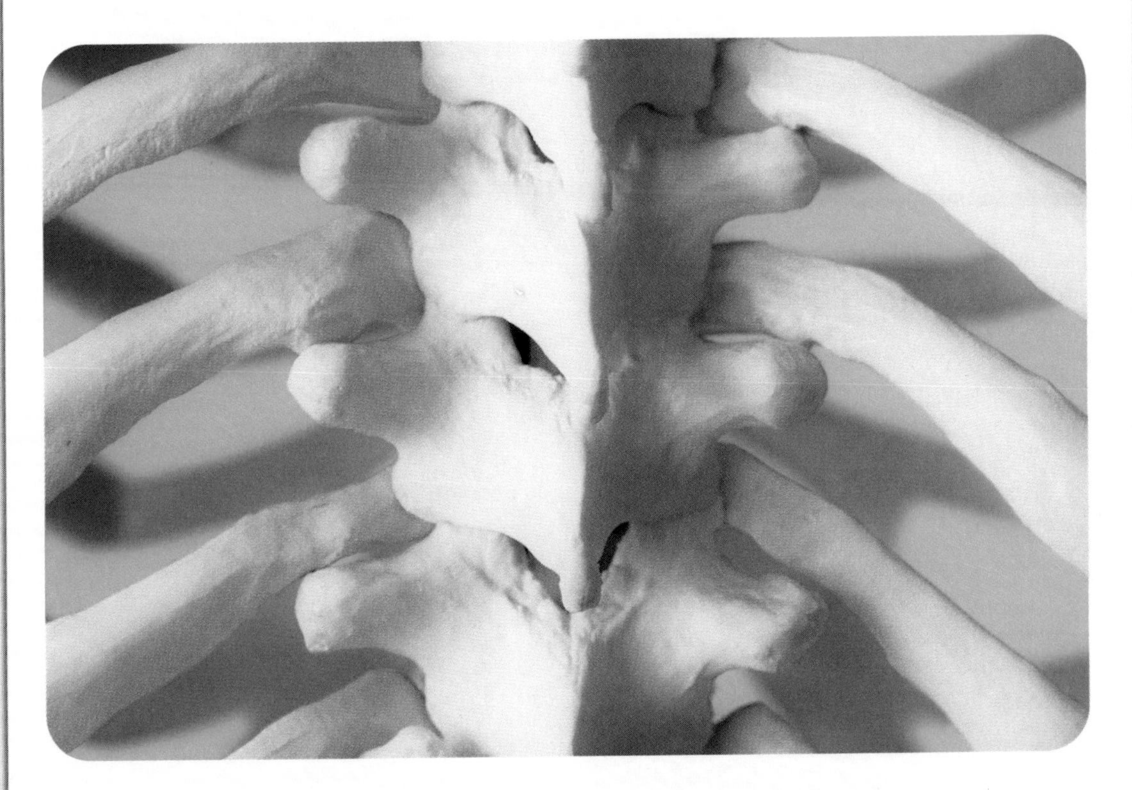

▲ What patterns do you see in the human spine and ribs?

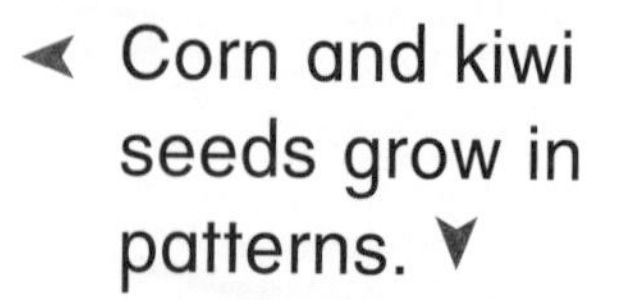

Corn and kiwi seeds grow in patterns.

Describe the patterns you see on the inside and outside of nautilus shells.

Mathematics ... Every Day

What patterns do you see in this cloth?

Pueblo Indians paint patterns like these on their pottery.

Hula dancers make patterns when they dance together.

You can see patterns in the honeycomb made by bees.

What patterns do you see in this spiral staircase?

Look around. What patterns can you find in nature? What patterns can you find that are made by people?

Number Stories

Number Stories and Situation Diagrams

Read It Together

Number stories are stories that use numbers. One way to solve number stories is to use diagrams.

Some number stories are about a total and its parts. You can use a **parts-and-total diagram** to help you solve them.

- There are 8 yellow crayons and 6 blue crayons. How many crayons are there in all?

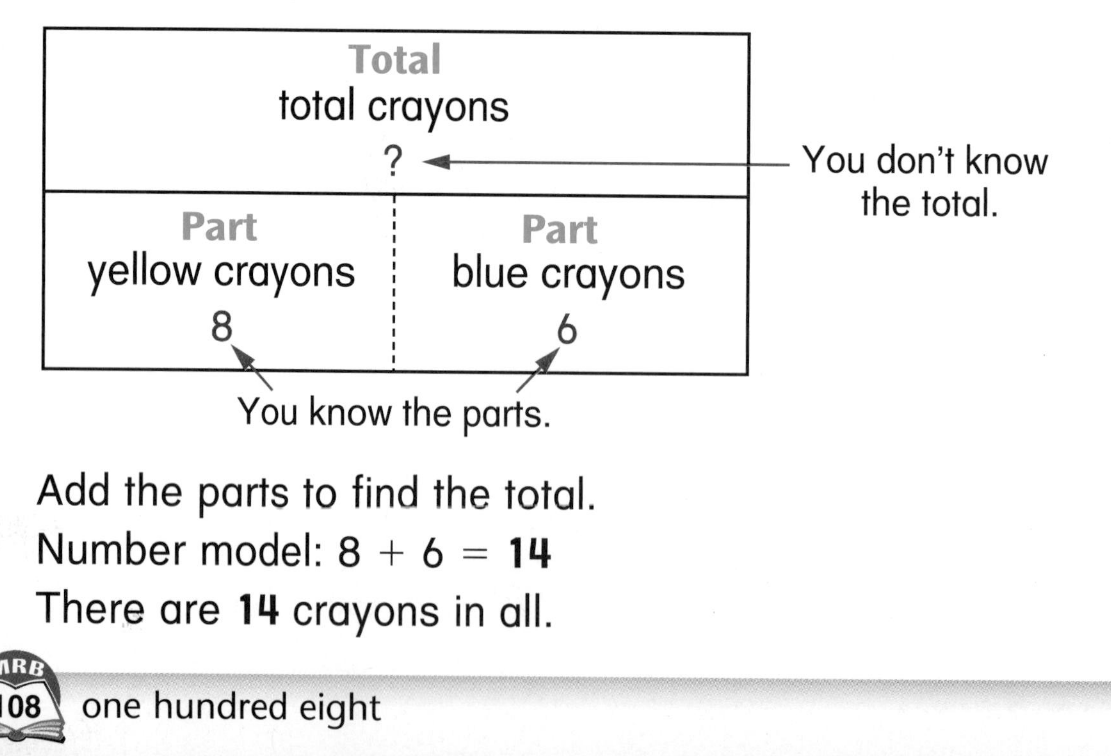

Add the parts to find the total.

Number model: 8 + 6 = **14**

There are **14** crayons in all.

Sometimes you need to find one of the parts.

- There are 24 children on a bus.
  9 children are girls. How many children are boys?

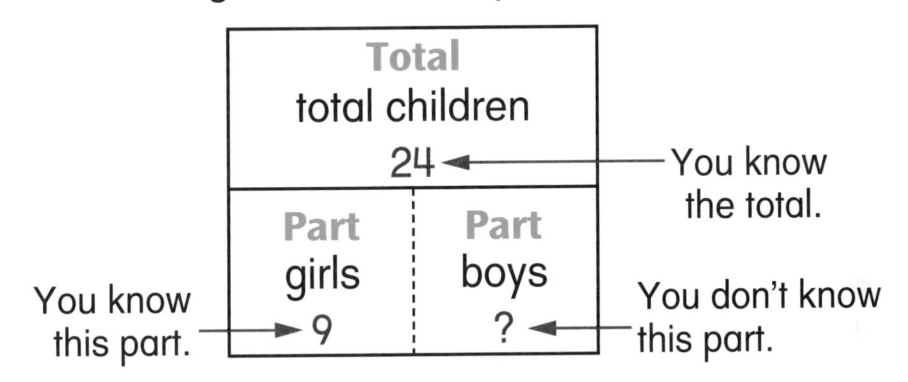

You can subtract to solve the problem.
Subtract the part you know from the total.
The answer is the other part.
Number model: 24 − 9 = **15** **15** boys on the bus

- You can also add up to solve the problem.
  Start with the part you know. Add up to the total.
  The amount you add up is the other part.

$$9 \xrightarrow{+1} 10 \xrightarrow{+10} 20 \xrightarrow{+4} 24$$

1 + 10 + 4 = **15**
Number model: 9 + **15** = 24 **15** boys on the bus

Some number stories are about comparisons. You can use a **comparison diagram** to help you solve them.

- Jim is 13 years old. Ron is 9 years old. How many years older is Jim than Ron?

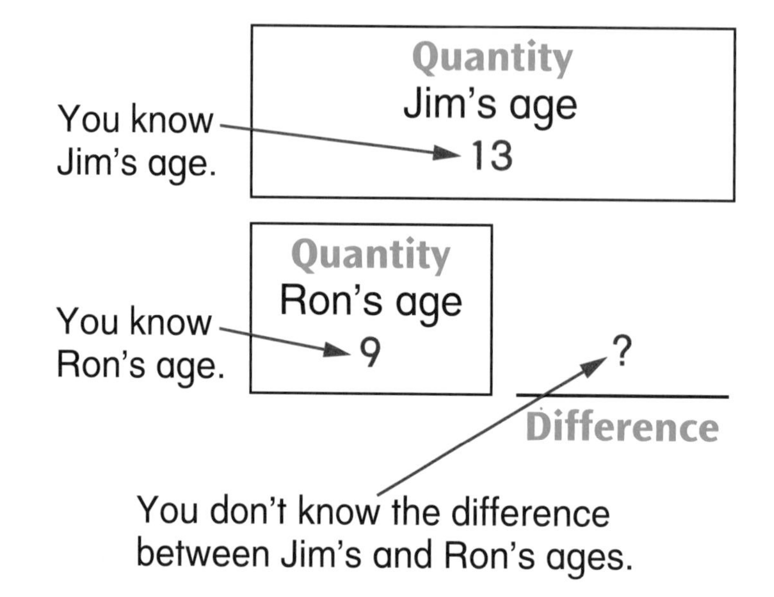

You can subtract to find the difference.
Start with the larger number.
Subtract the smaller number.
The answer is the difference.

Number model: 13 − 9 = **4**

Jim is **4** years older than Ron.

- You can also add up to find the difference.
  Start with the smaller number.
  Add up to the larger number.
  The amount you add up is the difference.

$$9 \xrightarrow{+1} 10 \xrightarrow{+3} 13$$

$1 + 3 = \mathbf{4}$

Number model: $9 + \mathbf{4} = 13$

Jim is **4** years older than Ron.

Tell your partner a comparison number story.

You can use a diagram to help you solve a number story about groups with equal numbers of objects.

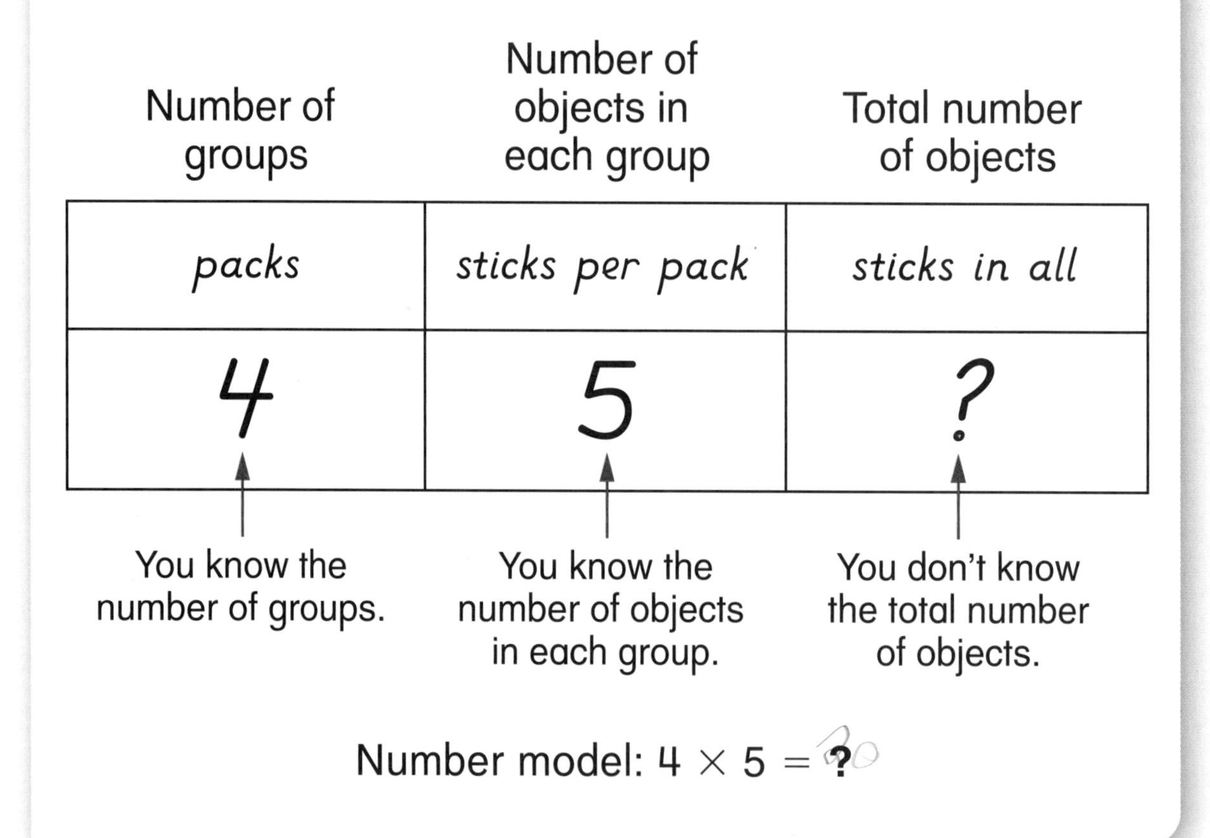

- Mia has 4 packs of gum.
  There are 5 sticks of gum in each pack.
  How many sticks of gum are there in all?

Number of groups	Number of objects in each group	Total number of objects
packs	*sticks per pack*	*sticks in all*
4	5	?
You know the number of groups.	You know the number of objects in each group.	You don't know the total number of objects.

Number model: $4 \times 5 = ?$

There are many ways to solve the problem.

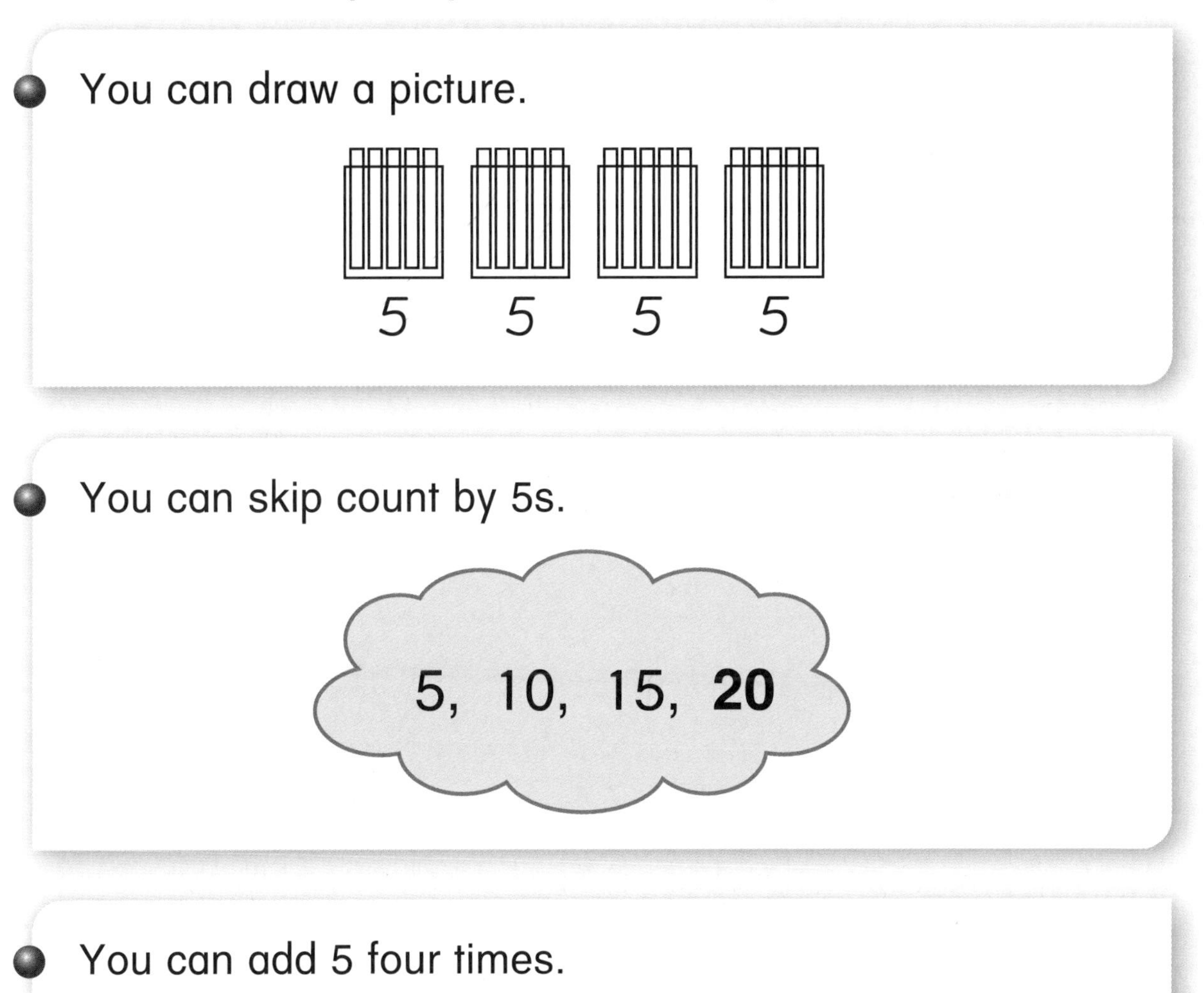

- You can draw a picture.

5 5 5 5

- You can skip count by 5s.

5, 10, 15, **20**

- You can add 5 four times.

$5 + 5 + 5 + 5 = \mathbf{20}$

Number model: $4 \times 5 = \mathbf{20}$

There are **20** sticks of gum in all.

Equal shares number stories are about dividing groups of objects into parts called **shares.** You can use a diagram to help you solve an equal share story.

- There are 24 marbles and 3 children.
  Each child should get the same number of marbles.
  How many marbles should each child get?

Number of shares	Number of objects in each share	Total number of objects
children	*marbles per child*	*marbles in all*
3	?	24
You know the number of shares.	You don't know the number of objects in each share.	You know the total number of objects.

Number models: $3 \times \mathbf{?} = 24$

$24 \div 3 = \mathbf{?}$

- One way to solve the problem is to draw a picture.

- Another way is to use counters. Deal out 24 counters among 3 equal shares to see how many counters are in each share.

Number models: $3 \times \mathbf{8} = 24$

$24 \div 3 = \mathbf{8}$

Each child has **8** marbles.

In change stories, the number you start with changes to more or changes to less. You can use a **change diagram** to help you solve a change-to-more story.

- Britney had 7 shells.
  She found 9 more shells.
  How many shells does Britney have in all?

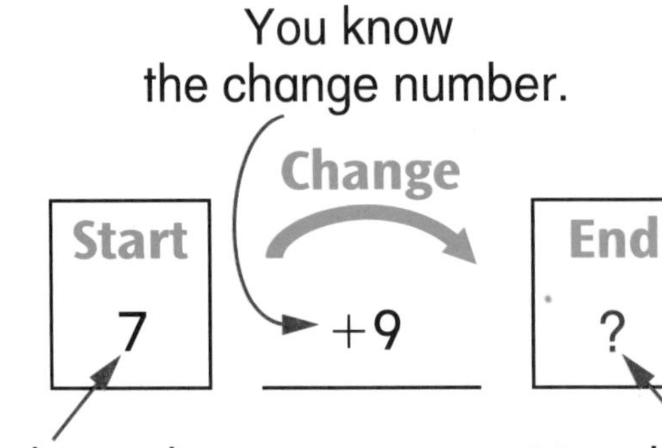

Number model: 7 + 9 = **16**

Britney has **16** shells in all.

You can use a change diagram to help you solve a change-to-less story.

- There are 15 children on a bus.
  6 children get off the bus.
  How many children are left on the bus?

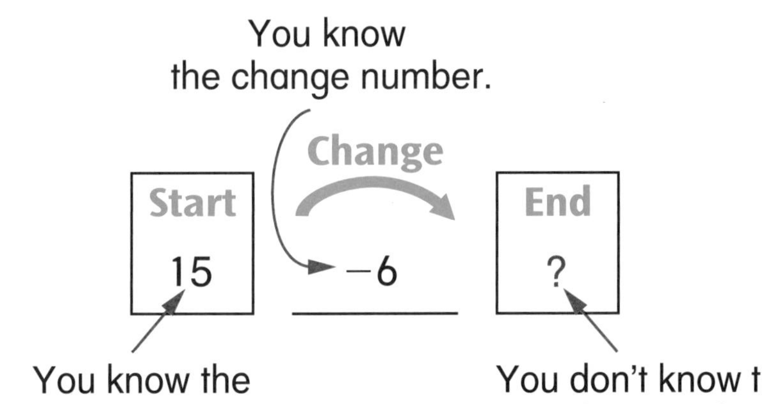

Number model: $15 - 6 = \mathbf{9}$

There are **9** children left on the bus.

Sometimes you need to find the change in a change story. You can use a change diagram to help you solve this kind of number story.

- The morning temperature was 50°F.
The afternoon temperature was 63°F.
What was the temperature change?

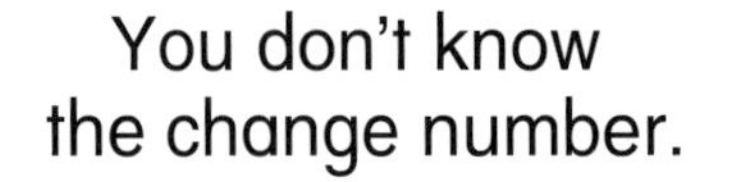

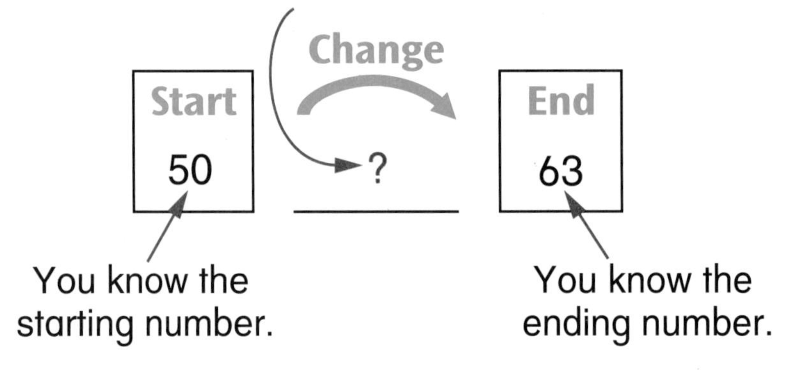

Number model: $50 + \mathbf{13} = 63$

The temperature change was $+\mathbf{13}$°F.

Try It Together

Take turns with a partner making up and solving number stories.

Games

Addition Spin

Materials ❑ 1 *Addition Spin* spinner
❑ 1 paper clip
❑ 1 pencil
❑ 1 calculator
❑ 2 sheets of paper

Players 2

Skill Mental addition

Object of the game To have the larger total.

Directions

1. Players take turns being the "Spinner" and the "Checker."

2. The Spinner uses a pencil and a paper clip to make a spinner.

3. The Spinner spins the paper clip.

4. The Spinner writes the number that the paper clip points to. If the paper clip points to more than one number, the Spinner writes the smaller number.

5. The Spinner spins a second time and writes the new number.
6. The Spinner adds the 2 numbers and writes the sum. The Checker checks the sum of the 2 numbers by using a calculator.
7. If the sum is correct, the Spinner circles it. If the sum is incorrect, the Spinner corrects it but does not circle it.
8. Players switch roles. They stop after they have each played 5 turns. Each player uses a calculator to find the total of his or her circled scores.
9. The player with the larger total wins.

Sam spins 5 and 25. He writes 30.
Mia spins 10 and 10. She writes 20.
Sam has the larger sum.

Addition Top-It

Materials ❑ number cards 0–10 (4 of each)

Players 2 to 4

Skill Addition facts 0 to 10

Object of the game To collect the most cards.

Directions

1. Shuffle the cards. Place the deck number-side down on the table.
2. Each player turns over 2 cards and calls out the sum of the numbers.
3. The player with the largest sum wins the round and takes all the cards.
4. In case of a tie for the largest sum, each tied player turns over 2 more cards and calls out the sum of the numbers. The player with the largest sum then takes all the cards from both plays.
5. The game ends when not enough cards are left for each player to have another turn.
6. The player with the most cards wins.

- Andrew turns over a 2 and a 3. He says, "2 plus 3 equals 5."

  Tanya turns over an 8 and a 9. She says, "8 plus 9 equals 17. 17 is more than 5. I take all four cards."

Another Way to Play

Use dominoes instead of cards.

Beat the Calculator

Materials ❑ number cards 0–9 (4 of each)
❑ 1 calculator

Players 3

Skill Mental addition

Object of the game To add numbers faster than a player using a calculator.

Directions

1. One player is the "Caller." A second player is the "Calculator." The third player is the "Brain."
2. Shuffle the cards. Place the deck number-side down on the table.
3. The Caller draws 2 cards from the number deck and asks for the sum of the numbers.

4. The Calculator solves the problem *with* a calculator. The Brain solves it *without* a calculator. The Caller decides who got the answer first.

5. The caller continues to draw 2 cards at a time from the number deck and to ask for the sum of the numbers.

6. Players trade roles every 10 turns or so.

- The Caller draws a 2 and a 9. The Caller says, "2 plus 9."

  The Brain and the Calculator each solve the problem.

  The Caller decides who got the answer first.

Another Way to Play

The Caller can choose problems from the Facts Table.

Before and After

Materials ❑ number cards 0–10 (4 of each)

Players 2

Skill Identifying numbers that are 1 less or 1 more than a given number

Object of the game To have fewer cards.

Directions

1. Shuffle the cards.
2. Deal 6 cards to each player.
3. Place 2 cards number-side up on the table.
4. Put the rest of the deck number-side down.
5. Players take turns. When it is your turn:
   - Look for any number in your hand that comes *before* or *after* one of the numbers on the table. Put it on top of the number. Play as many cards as you can.
   - Take as many cards as you need from the deck so that you have 6 cards again.

- If you can't play any cards when it is your turn, take 2 cards from the deck. Place them number-side up on top of the 2 numbers on the table. Try to play cards from your hand again. If you still can't, your turn is over.

6. The game is over when:
- All cards have been taken from the deck.
- No one can play any more cards.

7. The player holding fewer cards wins.

Sally gives 6 cards to Ricky and 6 cards to herself. Then she turns over a 9 card and a 2 card and places them on the table.

Ricky puts an 8 card on top of the 9 card and says, "8 is before 9." Then Ricky takes another card from the deck.

Coin Exchange

Materials ❑ 20 pennies, 10 nickels, 10 dimes, 2 quarters
❑ 2 six-sided dice
❑ 1 sheet of paper labeled "Bank"

Players 2

Skill Coin equivalencies

Object of the game To have more money.

Directions

1. Place all of the coins in the "Bank."
2. Players take turns. When it is your turn, roll both dice and collect from the bank the amount shown on the dice.
3. Whenever you can:
   - Exchange 5 pennies for a nickel in the bank.
   - Exchange 2 nickels, or 5 pennies and 1 nickel, for a dime.
   - Exchange a combination of nickels and dimes for a quarter.

4. The game ends when there are no more quarters in the bank.

5. The player with more money wins.

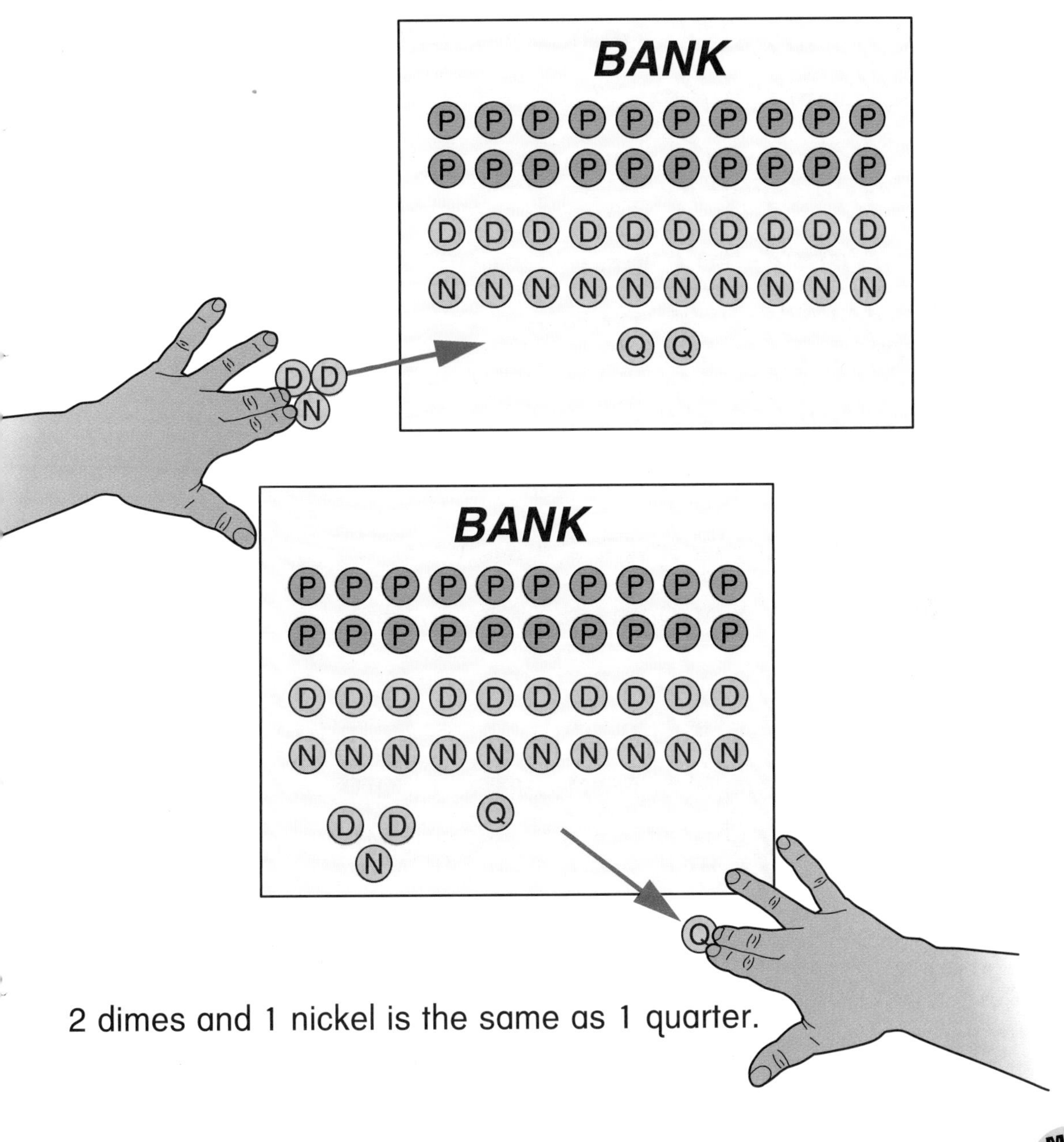

2 dimes and 1 nickel is the same as 1 quarter.

Difference Game

Materials
- ❑ number cards 1–10 (4 of each)
- ❑ 40 pennies
- ❑ 1 sheet of paper labeled "Bank"

Players 2

Skill Subtraction facts

Object of the game To take more pennies.

Directions

1. Shuffle the cards. Place the deck number-side down on the table.
2. Put 40 pennies in the bank.
3. To play a round, each player:
   - Takes 1 card from the top of the deck.
   - Takes the same number of pennies from the bank as the number shown on their card.
4. Find out how many more pennies one player has than the other. Pair as many pennies as you can.
5. The player with more pennies keeps the extra pennies. The rest of the pennies go back into the bank.

6. The game is over when there are not enough pennies in the bank to play another round.

7. The player with more pennies wins the game.

- Amy draws an 8. She takes 8 pennies from the bank.

John draws a 5. He takes 5 pennies from the bank.

Amy and John pair as many pennies as they can.

Amy has 3 more pennies than John. She keeps the 3 extra pennies and returns 5 of her pennies to the bank. John returns his 5 pennies to the bank.

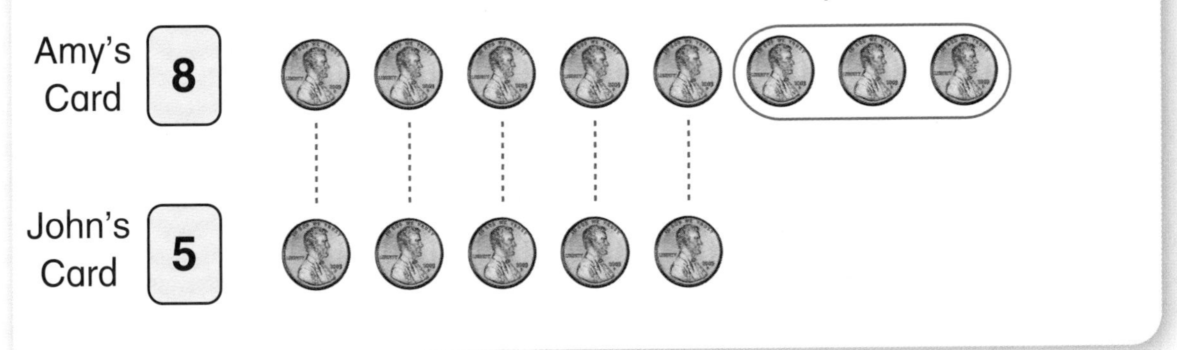

The Digit Game

Materials ❑ number cards 0–9 (4 of each)

Players 2

Skill Making and comparing numbers

Object of the game To collect more cards.

Directions

1. Shuffle the cards. Place the deck number-side down on the table.
2. Each player draws 2 cards from the deck and uses them to make the larger 2-digit number.
3. The player with the larger number takes all 4 cards.
4. The game is over when all of the cards have been used.
5. The player with more cards wins.

- Tina draws a 5 and a 3. She makes the number 53. Raul draws a 1 and a 4. He makes the number 41.

Tina

Tina's cards are a 5 and a 3.

Tina makes the number 53.

Raul

Raul's cards are a 1 and a 4.

Raul makes the number 41.

Tina's number is larger than Raul's number.
Tina takes all 4 cards.

Other Ways to Play

Make 3-Digit Numbers: Each player draws 3 cards and uses them to make the largest 3-digit number possible. Play continues as in the regular game.

Fewer Cards Win: The player with fewer cards wins.

Fact Extension Game

Materials
- ❑ number cards 0–9 (4 of each)
- ❑ 1 six-sided die
- ❑ 1 calculator
- ❑ 1 sheet of paper for each player

Players 2

Skill Finding sums of 2-digit numbers and multiples of 10

Object of the game To have the higher total.

Directions

1. Shuffle the cards. Place the deck number-side down on the table.
2. Each player draws 2 cards from the deck and makes the larger 2-digit number.
3. Players take turns rolling the die and making another 2-digit number by using the number on the die in the tens place and a zero in the ones place.
4. Each player adds his or her 2 numbers and records the sum on a sheet of paper.

5. After 4 rounds, players use a calculator to find the total of their 4 sums.

6. The player with the higher total wins.

- Anna draws a 3 and a 5. She makes the number 53. Then Anna rolls a 6. She makes the number 60.

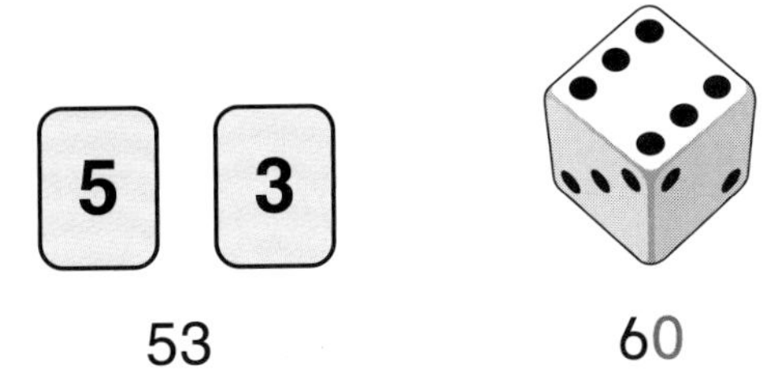

Anna finds the sum of her numbers.

53 + 60 = 113

Hit the Target

Materials ❑ 1 calculator
❑ 1 *Hit the Target* Record Sheet for each player

Players 2

Skill Finding differences between 2-digit numbers and multiples of ten

Object of the game To reach the target number.

Directions

1. Players agree on a 2-digit multiple of 10 as a *target number*. Each player records the target number on a Record Sheet.

2. Player 1 selects a *starting number* less than the target number and records it on Player 2's Record Sheet.

3. Player 2 enters the starting number into the calculator and tries to change the starting number to the target number by adding or subtracting on the calculator.

4. Player 2 continues adding or subtracting until the target number is reached.

5. Player 1 records each change and its result on Player 2's Record Sheet.

6. Players switch roles. Player 2 selects a starting number for Player 1. Player 2 fills in Player 1's Record Sheet as Player 1 operates the calculator.

7. The player who reaches the target number in fewer tries wins the round.

Hit the Target Record Sheet

Name ____________ Date ________ Time ________

Hit the Target **Record Sheet**

Round 1

Target number: ________

Starting Number	Change	Result	Change	Result	Change	Result

Round 2

Target number: ________

Starting Number	Change	Result	Change	Result	Change	Result

Round 3

Target number: ________

Starting Number	Change	Result	Change	Result	Change	Result

Round 4

Target number: ________

Starting Number	Change	Result	Change	Result	Change	Result

Name that Number

Materials ❑ number cards 0–20 (4 of each card 0–10, and 1 of each card 11–20)

Players 2 to 4 (the game is more fun when played by 3 or 4 players)

Skill Using addition and subtraction to name equivalent numbers

Object of the game To collect the most cards.

Directions

1. Shuffle the deck and place 5 cards number-side up on the table. Leave the rest of the deck number-side down. Then turn over the top card of the deck and lay it down next to the deck. The number on this card is the number to be named. Call this number the *target number*.

2. Players take turns. When it is your turn:
   - Try to name the target number by adding or subtracting the numbers on 2 or more of the 5 cards that are number-side up. A card may be used only once for each turn.

- If you can name the target number, take the cards you used to name it. Also take the target-number card. Then replace all the cards you took by drawing from the top of the deck.
- If you cannot name the target number, your turn is over. Turn over the top card of the deck and lay it down on the target-number pile. The number on this card is the new target number.

3. Play continues until all of the cards in the deck have been turned over. The player who has taken the most cards wins.

Mae and Joe take turns.

10 8

6

It is Mae's turn. The target number is 6.
Mae names the number with 12 − 4 − 2.
She also could have said 4 + 2 or 8 − 2.

Mae takes the 12, 4, 2, and 6 cards. Then she replaces them by drawing cards from the deck.

Now it is Joe's turn.

Number-Grid Difference

Materials
- ☐ number cards 0–9 (4 of each)
- ☐ 1 completed number grid
- ☐ 1 *Number-Grid Difference* Record Sheet
- ☐ 2 counters
- ☐ 1 calculator

Players 2

Skill Subtraction of 2-digit numbers using the number grid

Object of the game To have the lower sum.

Directions

1. Shuffle the cards. Place the deck number-side down on the table.

2. Take turns. When it is your turn:
   - Take 2 cards from the deck and use them to make a 2-digit number. Place a counter on the grid to mark your number.
   - Find the difference between your number and your partner's number.
   - This difference is your score for the turn. Write the 2 numbers and your score on the Record Sheet.

3. Continue playing until each player has taken 5 turns and recorded 5 scores.

4. Find the sum of your 5 scores. You may use a calculator to add.

5. The player with the lower sum wins the game.

- Ellie draws two 4s. She makes the number 44 and records it as her number on the Record Sheet.

  Carlos draws a 6 and a 3 and makes the number 63. Ellie records 63 on the Record Sheet. Then Ellie subtracts.

  $63 - 44 = 19$

  Ellie records 19 as the difference.

Number Grid Difference **Record Sheet**

Name ______ Date ______ Time ______

Number-Grid Difference Game Record Sheet

My Record Sheet

Round	My Number	My Partner's Number	Difference (Score)
1			
2			
3			
4			
5			
			Total ______

Round	My Number	My Partner's Number	Difference (Score)
1			
2			
3			
4			
5			
			Total ______

Number-Grid Game

Materials
- ❑ 1 number grid
- ❑ 1 six-sided die
- ❑ 1 game marker for each player

Players 2 or more

Skill Counting on the number grid

Object of the game To land on 110 with an exact roll.

Roll	Spaces
1	1 or 10
2	2 or 20
3	3
4	4
5	5
6	6

Directions

1. Players put their markers at 0 on the number grid.
2. Take turns. When it is your turn:
   - Roll the die.
   - Use the table to see how many spaces to move your marker.
   - Move your marker that many spaces.
3. Continue playing. The winner is the first player to land on 110 with an exact roll.

Number Grid

−9	−8	−7	−6	−5	−4	−3	−2	−1	0
1	2	3	4	5	6	7	8	9	10
11	12	13	14	15	16	17	18	19	20
21	22	23	24	25	26	27	28	29	30
31	32	33	34	35	36	37	38	39	40
41	42	43	44	45	46	47	48	49	50
51	52	53	54	55	56	57	58	59	60
61	62	63	64	65	66	67	68	69	70
71	72	73	74	75	76	77	78	79	80
81	82	83	84	85	86	87	88	89	90
91	92	93	94	95	96	97	98	99	100
101	102	103	104	105	106	107	108	109	110

One-Dollar Exchange

Materials ❑ 1 dollar, 20 dimes, 20 pennies
❑ 1 Place-Value Mat per player
❑ 2 six-sided dice
❑ 1 sheet of paper labeled “Bank”

Players 2

Skill Coin and bill equivalencies

Object of the game To make the exchange for a dollar.

Directions

1. Place all of the money in the “Bank.”

2. Players take turns. When it is your turn:

- Roll the dice and find the total number of dots.
- Take that number of cents from the bank and place the coins on the Place-Value Mat.
- If there are 10 or more pennies in the Pennies column, exchange 10 pennies for 1 dime. Place the dime in the Dimes column.
- If there are 10 or more dimes in the Dimes column, exchange 10 dimes for 1 dollar.

3. The winner is the first player to make the exchange for a dollar.

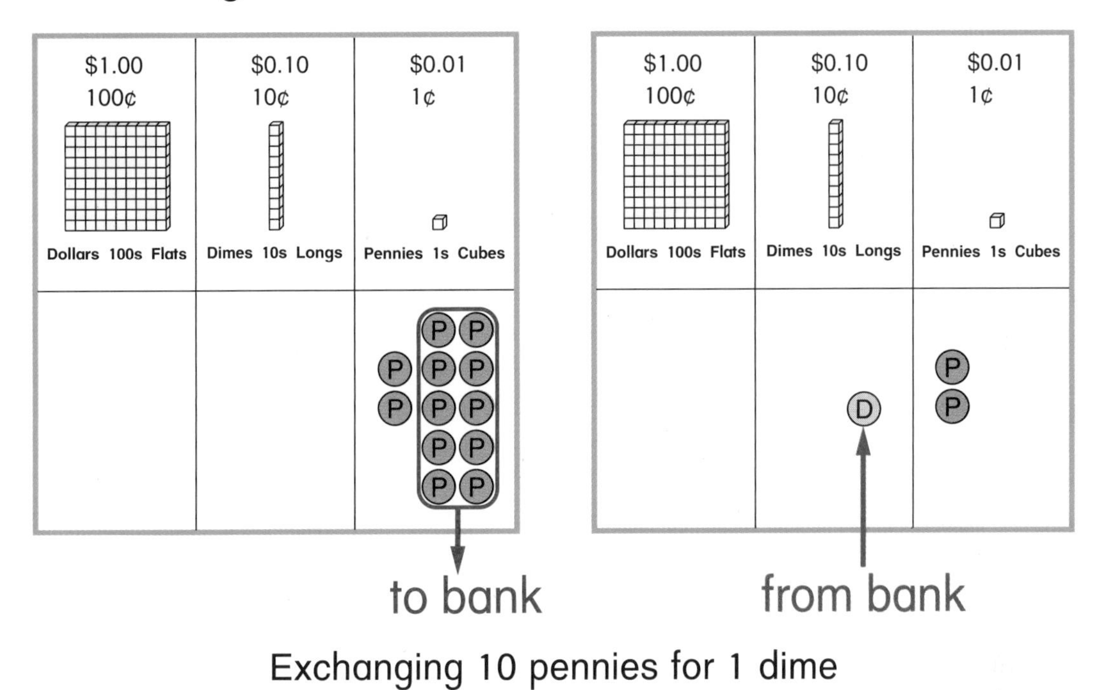

Exchanging 10 pennies for 1 dime

Penny Plate

Materials ❑ 10 pennies
❑ 1 small plastic plate

Players 2

Skill Sum-equals-ten facts

Object of the game To get 5 points.

Directions

1. Player 1:
   - Turns the plate upside-down.
   - Hides some of the pennies under the plate.
   - Puts the remaining pennies on top of the plate.

2. Player 2:
   - Counts the pennies on top of the plate.
   - Figures out how many pennies are hidden under the plate.

3. If the number is correct, Player 2 gets a point.

4. Players trade roles and repeat Steps 1 and 2.

5. Each player keeps a tally of their points.
   The first player to get 5 points is the winner.

Another Way to Play

Use a different number of pennies.

Subtraction Top-It

Materials ❑ number cards 0–10 (4 of each)

Players 2 to 4

Skill Subtraction facts 0 to 10

Object of the game To collect the most cards.

Directions

1. Shuffle the cards. Place the deck number-side down on the table.
2. Each player turns over 2 cards and subtracts the smaller number from the larger number.
3. The player with the largest difference wins the round and takes all the cards.
4. In case of a tie for the largest difference, each tied player turns over 2 more cards and calls out the difference of the numbers. The player with the largest difference then takes all the cards from both plays.
5. The game ends when not enough cards are left for each player to have another turn.

6. The player with the most cards wins.

- Ari turns over a 7 and a 5. He says,
  "7 minus 5 equals 2."
  Lola turns over a 9 and a 6. She says,
  "9 minus 6 equals 3. 3 is more than 2.
  I take all four cards."

Another Way to Play

Use dominoes instead of cards.

3, 2, 1 Game

Materials ❑ 1 sheet of paper
❑ 1 pencil for each player

Players 2

Skill Mental subtraction skills

Object of the game To reach exactly 0.

Directions

1. Write 21 at the top of the sheet of paper.
2. Players take turns. When it is your turn, subtract 1, 2, or 3 from the last number written on the paper.
3. The first player who subtracts and gets 0 as the answer wins the game.

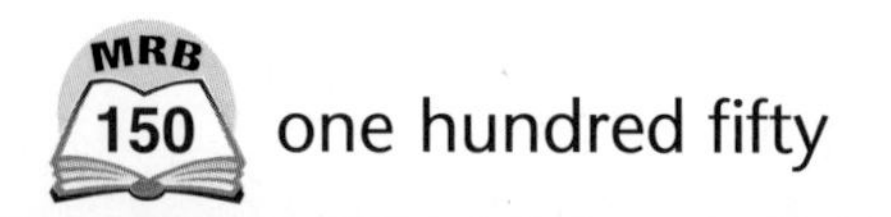

21
−2
19
−3
16
−3
13
−1
12

Start with 21.
Player 1 subtracts 2.

Player 2 subtracts 3.

Player 1 subtracts 3.

Player 2 subtracts 1.

Each player has taken 2 turns.
The game continues until a player subtracts and gets 0 as the answer.

Time Match

Materials ❑ 12 *Time Match* analog clock cards
❑ 12 *Time Match* digital clock cards

Players 2 or 3

Skill Telling time

Object of the game To match the most cards.

Directions

1. Shuffle the cards. Place all 24 cards clock-side down on the table in a 4-by-6 array.
2. Players take turns. When it is your turn:
   - Turn over 2 cards.
   - If the cards match, take them. Your turn is over.
   - If the cards don't match, turn them back over so that they are clock-side down again in the same position. Your turn is over.
3. When all of the cards have been collected, the player with the most matches wins.

- John turned over 2 cards. They both show 3:00. The cards match, so John took them.

  Yoko turned over 2 cards. One card shows 6:00. The other card shows 5:15. The cards do not match, so Yoko will turn them back over so that they are clock-side down again in the same position.

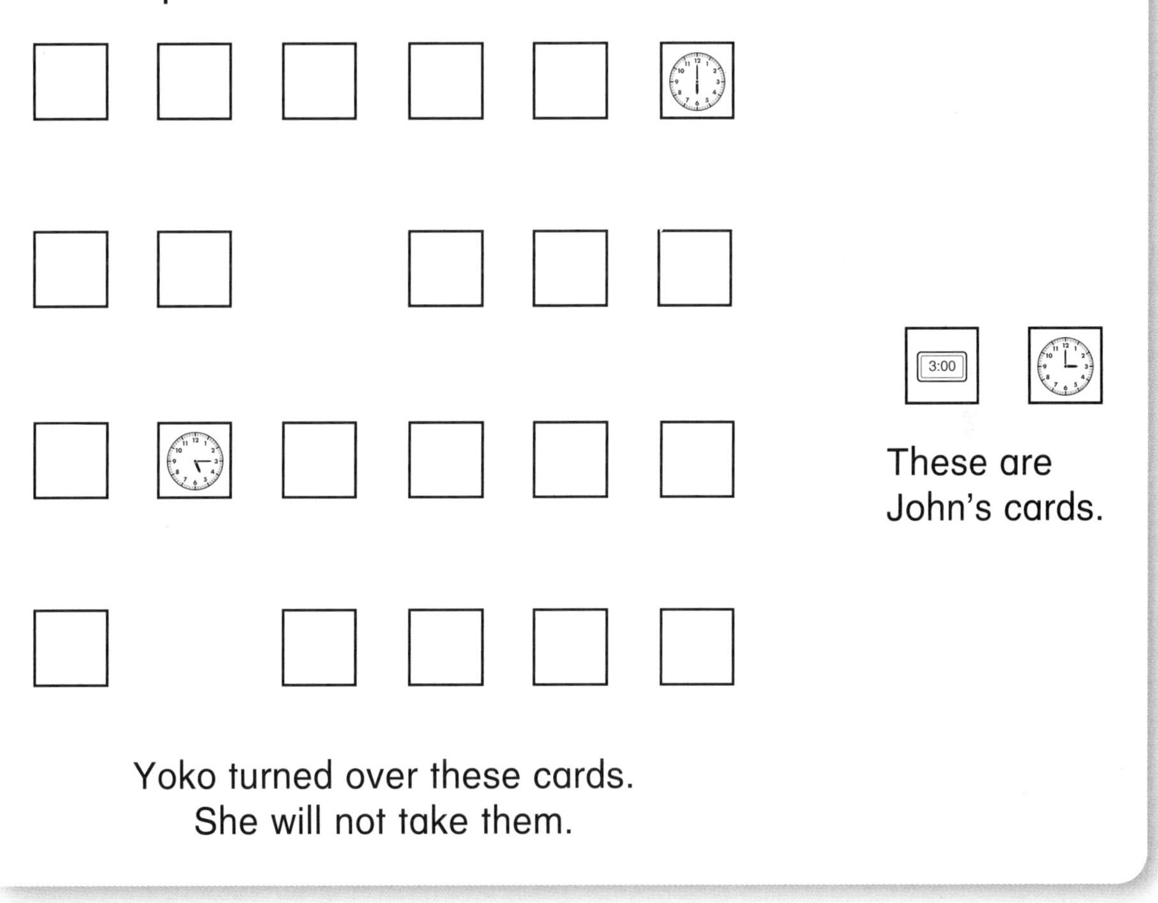

Top-It

Materials ❑ number cards 0–10 (4 of each)

Players 2

Skill Comparing numbers

Object of the game To collect more cards.

Directions

1. Shuffle the cards. Place the deck number-side down on the table.
2. Each player turns over 1 card and says the number on it.
3. The player with the larger number takes both cards. If both cards show the same number, each player turns over another card. The player with the larger number then takes all 4 cards for that round.

4. The game is over when all of the cards have been turned over.

5. The player with more cards wins.

- Pam turns over a 4. She says, "4."

  Mark turns over a 6. He says, "6.
  6 is larger than 4, so I take both cards."

Another Way to Play

Use dominoes instead of cards.

Tric-Trac

Materials
- ❑ 2 six-sided dice
- ❑ 20 pennies
- ❑ 1 *Tric-Trac* Game Mat for each player

Players 2

Skill Addition facts 0-10

Object of the game To have the lower sum.

Directions

1. Cover the empty circles on your game mat with pennies.
2. Take turns. When it is your turn:
   - Roll the dice. Find the total number of dots. This is your sum.
   - Move 1 of your pennies and cover your sum on your game mat.

OR

- Move 2 or more of your pennies and cover any numbers that can be added together to equal your sum.

3. Play continues until no more numbers can be covered on your game mat. Your partner may continue playing, even after you are finished.

4. The game is over when neither player can cover any more numbers on his or her game mat.

5. Find the sum of your uncovered numbers. The player with the lower sum wins.

Set-up *Tric-Trac* Game Mat

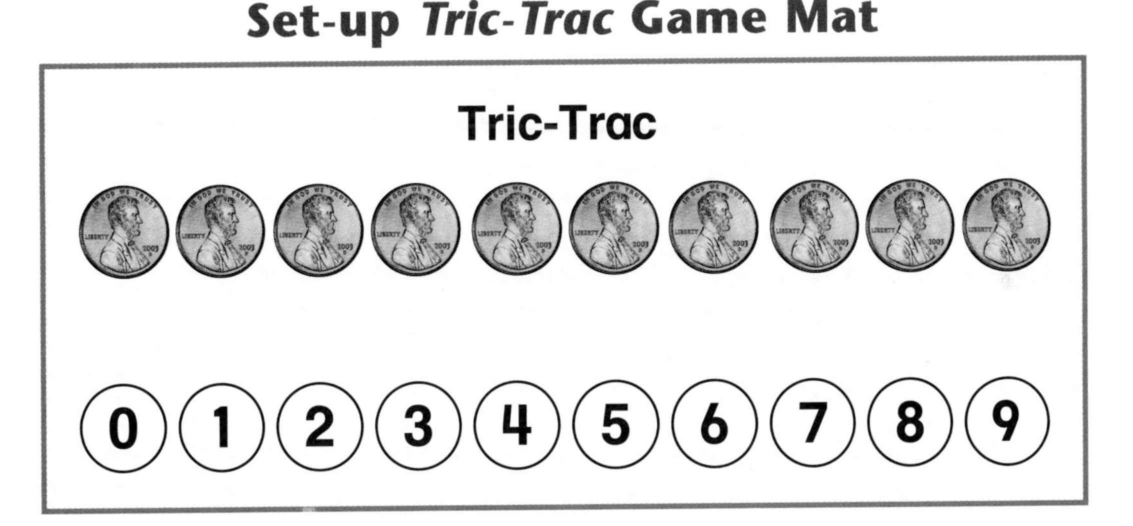

Turn to page 158 to see an example.

- David rolled a 1 and a 2.
David's sum is 3.

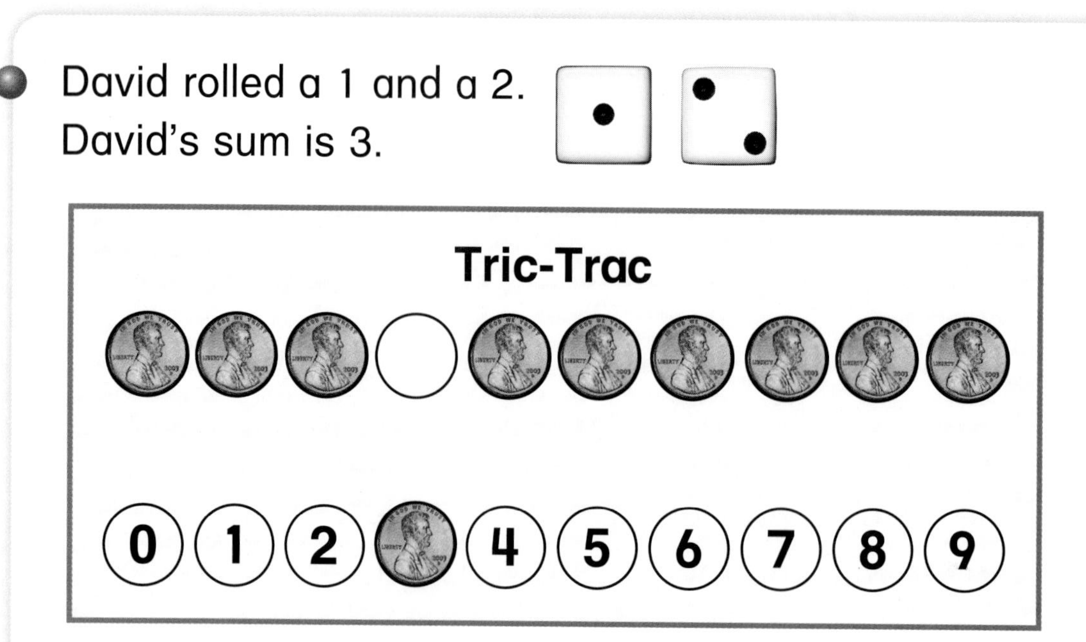

This is one way David can cover his sum.

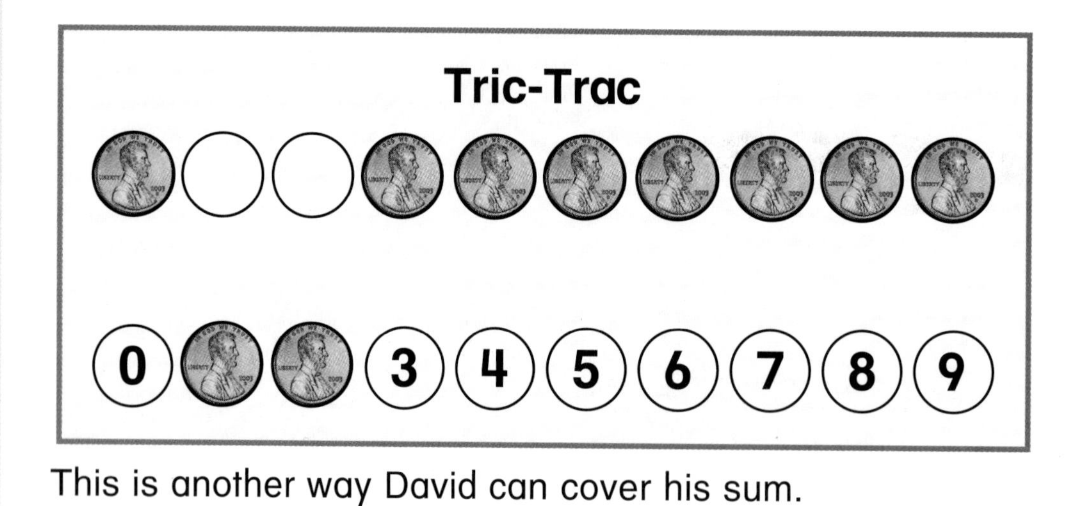

This is another way David can cover his sum.

Calculators

Calculators

Read It Together

A **calculator** is a tool that can help you do many things. You can use it to count, add, subtract, multiply, and divide. Not all calculators are alike.

- Here is one type of calculator.

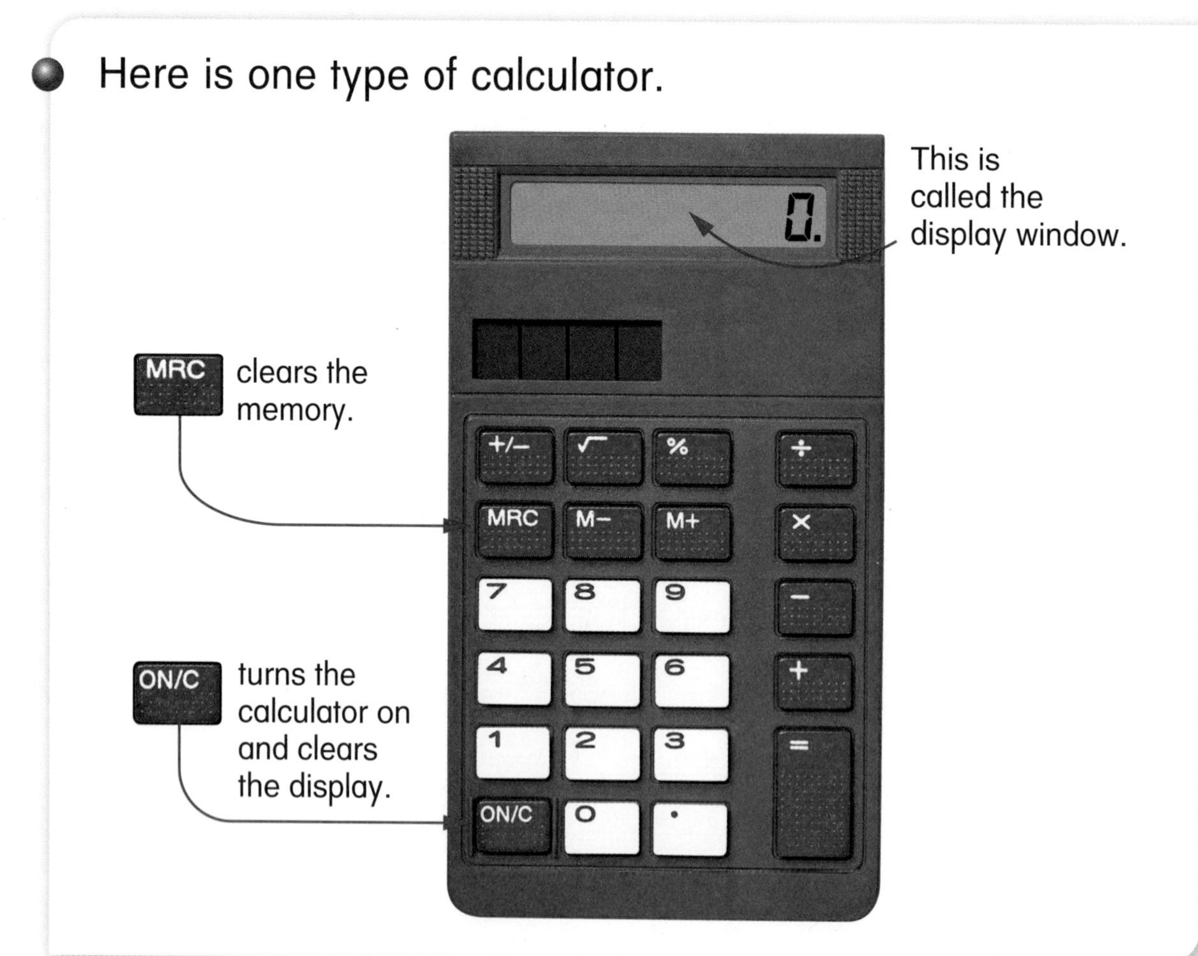

- Here is another type of calculator.

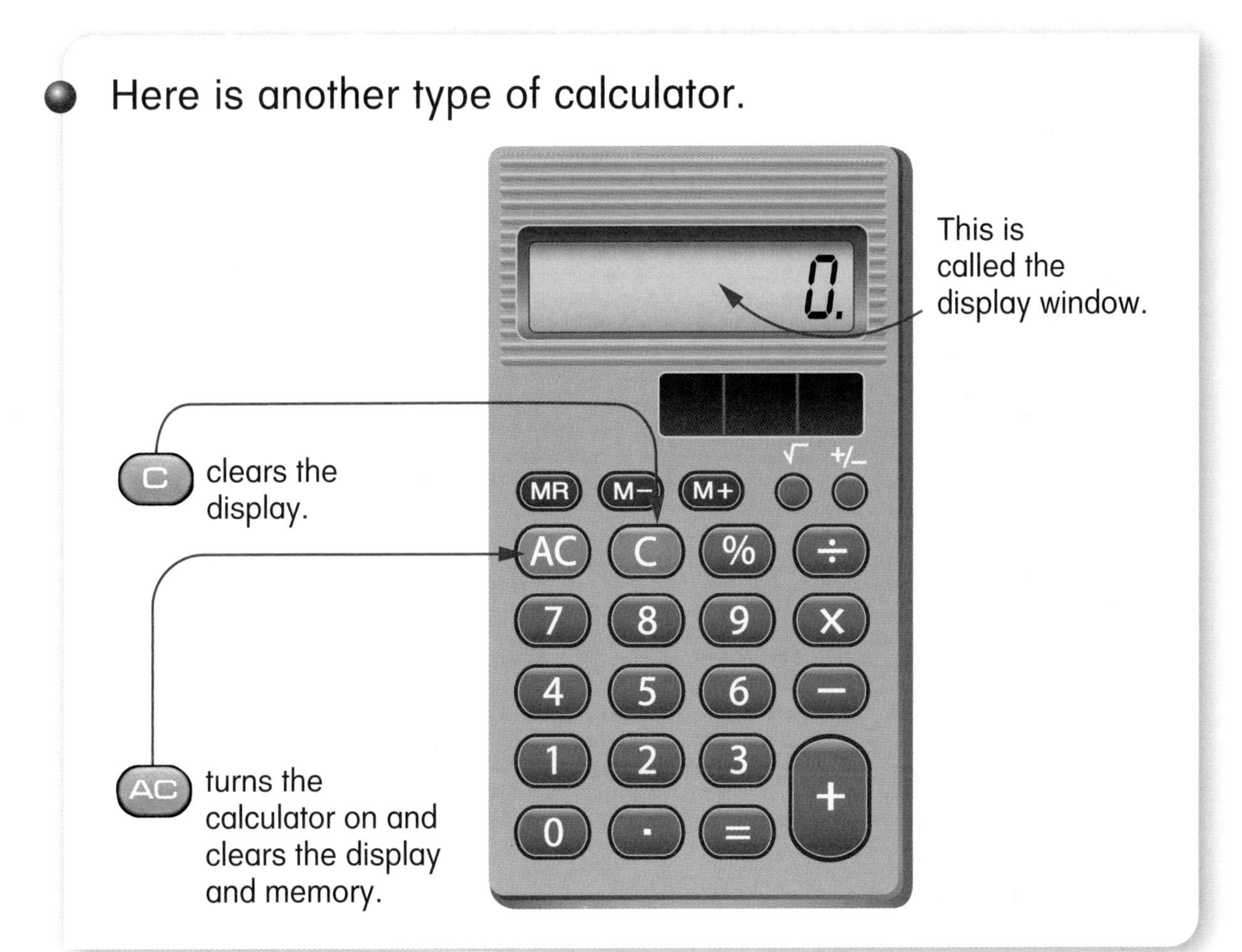

Did You Know?

Blaise Pascal, a famous mathematician, invented a calculator in 1642 at age 14.

You can **skip count** up or back on a calculator.

Use [calculator] to skip count. Start at 1. Count up by 2s.

Program the calculator	Keys to Press	Display
Clear the display.	[ON/C]	0.
Enter the starting number.	1	1.
Tell it to count up by a number.	[+] 2	2.
Tell it to skip to the next number.	[=]	3.
Tell it to skip to the next number.	[=]	5.
Tell it to skip to the next number.	[=]	7.

You can count back by pressing [–] instead of [+].

Note

When skip counting beginning with a negative number, it is necessary to enter the start number and then press the [+/–] key to make the number negative.

- Use [calculator] to skip count. Start at 1. Count up by 2s.

Program the calculator	Keys to Press	Display
Clear the display.	(AC)	0.
Tell it to count up by a number.	2 (+) (+)	K 2.+
Enter the starting number.	1	K 1.+
Tell it to skip to the next number.	(=)	K 3.+
Tell it to skip to the next number.	(=)	K 5.+
Tell it to skip to the next number.	(=)	K 7.+

You can count back by pressing (−) (−) instead of (+) (+).

Note

Always clear the calculator display before you start something new.

The keys for the **basic operations** are on every calculator.

- These tables show addition and subtraction problems.

Operation	Problem	Key Sequence	Display
Addition	2 + 4	2 [+] 4 [=]	6.
Addition	23 + 19	23 [+] 19 [=]	42.

Operation	Problem	Key Sequence	Display
Subtraction	10 − 7	10 [−] 7 [=]	3.
Subtraction	51 − 16	51 [−] 16 [=]	35.

Note

Ask yourself if the number in the display makes sense. This helps you know if you pressed a wrong key or made another mistake.

- These tables show multiplication and division problems.

Operation	Problem	Key Sequence	Display
Multiplication	2×5	2 (×) 5 (=)	10.
Multiplication	6×14	6 (×) 14 (=)	84.

Operation	Problem	Key Sequence	Display
Division	$10 \div 2$	10 (÷) 2 (=)	5.
Division	$92 \div 4$	92 (÷) 4 (=)	23.

Try It Together

Play *Beat the Calculator* on pages 124–125.

Index

D

E

F

G

MRB
168

O

P

R

S

Photo Credits

©Ablestock, p. 56; ©James L. Amos/Corbis, p. 19 *top*; ©Lester V. Bergman/Corbis, p. 19 *bottom right*; ©Jonathan Blair/Corbis, p. 74 *top*; ©Brand X Pictures, pp. 64 *second from top*, 70 *bottom right*; ©Corbis, p. 104 *top right*; ©DK Limited/Corbis, p. 104 *bottom left*; ©Getty Images, pp. 3 *bottom right*, 14, 57 *bottom*, 64 *fourth from top*, 70 *top right*, 93, 96; ©Pascal Goetgheluck/Science Photo Library, p. 74 *bottom left*; ©David Gray/Reuters/Corbis, p. 18 *top*; ©Hemera Technologies/Alamy Images, p. 70 *bottom left*; ©Peter Johnson/Corbis, p. 75 *center*; ©Jupiter Images, p. 70 *top left*; ©Wolfgang Kaehler/Corbis, p. 74 *center*; ©Frans Lanting/Minden Pictures, p. 75 *top*; ©Robert Lewelyn/Zefa/Corbis, p. 103; ©Jim McDonald/Corbis, p. 75 *bottom*; ©Joe McDonald/Corbis, p. 18, *bottom*; p. 74 *bottom right*; ©Mary Ann McDonald/Corbis, p. 18 *center*; ©John Morrison, pp. 2, 3 *top right*, 4 *top right*, 57 *top left*, 64 *top*, 146; ©NASA, p. 85; ©Photo Cuisine/Corbis, p. 104, *top left*; ©Photos.com, pp. 4 *center*, 64 *third from top*; ©Tom Prettyman/Photo Edit, p. 17; ©Linda Richardson/Corbis, p. 20; ©Kevin Shaefer/Corbis, p. 76; ©Ariel Skelley/Corbis, p. 105 *top left*. ©Joel Stettenheim/Corbis, p. 19 *bottom left*; ©Josh Westrich/Zefa/Corbis, p. 104 *bottom right*; ©Staffan Widstrand/Corbis, p. 73.